"Arora offers us another invitation, which is a refresh[ing antidote to the] breathlessness of many studies of the new technologies[. Arora invites us] to slow down, to pause, to contemplate our surroundings, to smell a possibly political rose. That she finds this potential in the very heart of digitality is one of the many surprises of this thoughtful and wide-ranging book."

– From the Foreword by Arjun Appadurai,
Paulette Goddard Professor of Media, Culture,
and Communication, New York University

"This is a brilliant navigation of worlds that are not usually brought in conversation: digital space and thick situated struggles engaged in claim-making in the urban sphere. Payal Arora has deep knowledge and experience of both these worlds. Out of this encounter comes a concept the author deploys in diverse ways to mark digital space: the leisure commons."

– Saskia Sassen, Columbia University
and author of *Expulsions: Brutality
and Complexity in the Global Economy*

"In this engaging volume, Arora applies the rich metaphor of the public park to explicate the many ways in which net-based technologies facilitate, but also converge activities of a social, political, cultural and economic nature. Technology as architecture invites, amplifies, but also conceals or discourages. It disrupts and it sustains our daily endeavors into sociality, work, play and fantasy. Arora uses the metaphor of public parks to tell the story of how digital media support us through our daily lives. Through lively writing and layers of intriguing analogies, she compels the reader to think with her, as she explores what technology does to space. Arora lays out an intriguing vision of online environments as technology supported meta-parks that facilitate not just limitless connection, but, better living."

– Zizi Papacharissi, Professor and
Head of Communications,
University of Illinois at Chicago

"Payal Arora offers the insight that social media are the latest chapter in a long history of spaces including city parks, walled gardens, office parks, fantasy theme parks and other semi-public, leisure-oriented environments. By framing new technological trends in terms of a 'leisure commons,' her work fills a gap that remained between the spatial metaphors that have proven helpful to make sense of new technologies, and a nuanced realization of how thoroughly leisure practices have permeated daily life."

– Paul C. Adams, Associate Professor
of Geography and Director of Urban
Studies, University of Texas at Austin

The Leisure Commons

There is much excitement about Web 2.0 as an unprecedented, novel, community-building space for experiencing, producing, and consuming leisure, particularly through social network sites. What is needed is a perspective that is invested in neither a utopian or dystopian posture but sees historical continuity to this cyberleisure geography. This book investigates the digital public sphere by drawing parallels to another leisure space that shares its rhetoric of being open, democratic, and free for all: the urban park. It makes the case that the history and politics of public parks as an urban commons provides fresh insight into contemporary debates on corporatization, democratization and privatization of the digital commons. This book takes the reader on a metaphorical journey through multiple forms of public parks such as Protest Parks, Walled Gardens, Corporate Parks, Fantasy Parks, and Global Parks, addressing issues such as virtual activism, online privacy/surveillance, digital labor, branding, and globalization of digital networks. Ranging from the 19th century British factory garden to Tokyo Disneyland, this book offers numerous spatial metaphors to bring to life aspects of new media spaces. Readers looking for an interdisciplinary, historical and spatial approach to staid Web 2.0 discourses will undoubtedly benefit from this text.

Payal Arora is an Assistant Professor in the Department of Media and Communication at Erasmus University Rotterdam. She is the author of *Dot Com Mantra: Social Computing in the Central Himalayas* (Ashgate, 2010) and winner of the 2010 *Social Informatics Best Paper Award* from ASIS&T.

Routledge Studies in Science, Technology and Society

The Leisure Commons

A Spatial History of Web 2.0

Payal Arora

Routledge
Taylor & Francis Group
NEW YORK   LONDON

First published 2014
by Routledge

Published 2016 by Routledge

711 Third Avenue, New York, NY 10017

and by Routledge
2 Park Square, Milton Park, Abingdon, Oxon OX14 4RN

*Routledge is an imprint of the Taylor & Francis Group,
an informa business*

First issued in paperback 2016

Library of Congress Cataloging-in-Publication Data
Arora, Payal.
   The leisure commons : a spatial history of Web 2.0 / Payal Arora.
      pages cm. — (Routledge studies in science, technology and society
; 27)
   Includes bibliographical references and index.
   1.   Web 2.0. 2.   Internet—Social aspects. 3.   Social networks.
   4.   Space—Social aspects.   I. Title.
   TK5105.88817.A76 2014
   025.042—dc23
   2013047615

ISBN13: 978-1-138-68804-9 (pbk)
ISBN13: 978-0-415-88711-3 (hbk)

Typeset in Sabon
by IBT Global.

To Panchmi

Always by my side

Contents

Figures

Foreword

This exhilarating book does the rarest of things, which is to render an inert metaphor live and illuminating. The metaphor in question is the one of the 'park' or 'garden' as a common space for leisure as a shared or public activity. This humble element of our geographical worlds is itself an interesting innovation, which is not confined to one culture or society, though it has been more developed at some times and places than in others.

It turns out that the World Wide Web, in its most recent incarnation, is also seen as a space of play, of leisure, and of fun-in-common, both by its students and by its inhabitants. Payal Arora makes an eloquent case for why this parallel between cyberspace and lived geographical space is neither an accident nor a minor overlap.

Her book takes this metaphor, and all metaphors, seriously, and in this inclination she is in good company, as many scholars of the social life of language have shown. It turns out that the idea of the cyber-commons as a kind of walled garden is an active part of the metaphorical life of many players in cyberspace and the exploration of this resemblance turns out to have many implications for some of the grand issues of our times, such as leisure, politics, privacy, and community.

Indeed the idea that cyberspaces have some of the characteristics of parks and gardens, and that parks and gardens have frequently been the site of struggles about who can enjoy them and who cannot, also explains how park-like cyberspaces also return to the social, in the form of fantasy parks, corporate parks, and other sorts of creative enclosures. And as enclosures invariably raise the questions of exclusion and inequality, Arora does not shy away from taking on the question of what it means as globalization proliferates its presence in part through various forms of cyber-leisure and cyber-pleasure.

And as she takes us through the gardens and parks of our new technological world, Arora offers us another invitation, which is a refreshing departure from the breathlessness of many studies of the new technologies, and that is the chance to slow down, to pause, to contemplate our surroundings, to smell a possibly political rose. That she finds this potential in the very heart of digitality is one of the many surprises of this thoughtful and wide-ranging book.

Arjun Appadurai
New York University
September 2013

Preface

The idea for this book emerged way back during my doctoral years in 2005. While strolling through Riverside Park in New York City, it came to me how seemingly unregulated a park space appears to be and, yet, people organize themselves within these leisure terrains in a myriad of ways. True to a typical doctoral student, I deviated from my doctoral thesis by writing a paper on 'Online Social Sites as Virtual Parks: An Investigation into Leisure Online and Offline' (this was later published in the *Information Society Journal*). I found that by drawing parallels between the space of Web 2.0 and urban parks, it inspired friends, family, and colleagues to participate in this discussion. Through the disarming metaphor of public parks, it made the web familiar to those who were not traditionally inclined to talk about new media. Before I knew it, I was engaging in conversations on how the Speakers' Corner in Hyde Park resembles online protest space, how it is not a coincidence that 'walled gardens' is used as a metaphor for the privatization of the web, and that amusement parks have much in common with implicit branding within digital gaming sites.

I was encouraged through challenging conversations with Todd Gitlin and Robert McClintock from Columbia University, which pushed me to approach these playful ideas on virtual parks in a more rigorous way. My discussion with Arjun Appadurai opened up further parallels of equating the flâneur to the digital user, browsing leisurely through the web. In fact, I am particularly grateful to Professor Appadurai for his sustained mentorship and support over the years as I embarked on taking this topic seriously and plunged into the writing of this book.

In 2012, I was fortunate to have received a generous grant for this book project through the Erasmus University Rotterdam Fellowship. I am thankful to Filip Vermeylen for his feedback in this process. This grant has allowed me to immerse myself in this topic wholeheartedly and participate in conferences across disciplines, for which I will always be grateful. I found that to realize this parallel, I needed to delve into unfamiliar disciplines, such as geography, history, and law. While indeed daunting to get out of the familiar terrain of communication studies, it has been exciting to encounter rich conversations in multiple fields, reinforcing interdisciplinary studies as a must for innovative thinking.

Over the last three years, I have presented on this subject at 20 conferences across disciplines, including at the American Sociological Association, European Sociological Association, International Communication Association, Society for Social Studies of Science, Amsterdam Privacy Conference, IIS World Congress, International Association for Media and Communication Research, and the Technology, Knowledge, and Society Conference. I am deeply indebted to the numerous blind reviewers and audiences at these conferences for their feedback and thought-provoking comments and suggestions.

It has been heartening to see this project gain traction over the years. This has manifested in invitations to speak at various prestigious events, including the Frontiers of New Media Symposium at the University of Utah, the Labor, Knowledge, and Leisure in Postindustrial Society conference in Moscow, and the NSF and Cornell University–sponsored symposium on Rethinking the Culture of Busyness and IT in Seattle. These opportunities have undoubtedly enriched my argument.

It is evident that all intellectual endeavors are deeply contingent on a nurturing professional and personal environment. This book is no different. I find myself fortunate to have an extremely supportive department chair and role model—Susanne Janssen. She has time and again gone out of her way to enable me in this journey. The faculty and the university at large have been deeply accommodating to my needs and interests and have undoubtedly made this pursuit easier for me. Yahya Kamalipour has been a wonderful support and in general has been there for me over the years as a mentor and a friend. Saskia Sassen, Julie Cohen, Helen Nissenbaum, Paul Adams, Zizi Papacharissi, Lori Kendall, Martin Dodge, Dan Hunter, Jonathan Taylor, and Manual Castells, among others, have been truly inspirational in the writing of this book.

It needs to be noted that without the strong editorial support of Janelle Ward, who was the first to read the entire text and edit it closely, this would have been a weaker manuscript. In this line, I am thankful to the Routledge editor Max Novick for his prompt responses to my ongoing queries and his overall professionalism in this process. Additionally, I am grateful to my research assistant, Jessica Verboom, a true asset to any academic engaged in such a scholarly pursuit. Besides being a gifted young scholar in her own right, she has patiently worked on this book with the referencing, formatting, images, and other details. This has, of course, enhanced the quality of this manuscript. I would also like to thank my research intern, Savo Bojovic, for his contribution as he gathered and reviewed sources on digital gaming that enriched the discussion on fantasy parks.

On a personal level, I gained much strength from my close friendships with Janelle Ward, Katalin Kabat-Ryan, Tom Ryan, Jennifer Turner, Leigh

Graham, Karen Kaun, Padraig Tobin, Bernadette Kester, Klazien de Vries, Filip Vermeylen, and John van Male. My partner, René König, has been an indelible support throughout this whole process and has patiently listened to my ramblings on this subject over the years. I count myself truly fortunate to have in my mother and father two deeply unobjective and tireless cheerleaders of my endeavors. Last but not least, this book is dedicated to my sister, Panchmi Arora, who is my best friend and ardent believer in my capabilities.

1   Introduction

The Internet has matured. It is now characterized by a new generation of websites popularly termed as Web 2.0. The nature of this transformation is predominantly social versus technical in nature, and is marked by the rise of social network sites and user-generated content. In particular, Web 2.0 is defined by its leisure properties. These leisure properties, this book will argue, are by no means completely novel. In the coming pages, we will explore how these leisure properties are deeply rooted in historical, socioeconomic, and cultural spaces and intrinsically tied to offline practices. In essence, to understand the nature of cyberleisure spaces, we need to examine closely their offline–online, transnational–transcultural, and historic–contemporary relationships.

This book proposes to use the metaphor of 'public parks' and its multiple forms to illustrate different dimensions of the digital commons. This metaphorical tool is used as a critical and comprehensive instrument of analysis. It is used to make the argument that public parks share the rhetoric of Web 2.0 spaces—that of being open, democratic, non-utilitarian, and free for all. However, rhetoric confronts reality that always comes with a rich and contentious historical struggle. By revealing the spectrum of tensions in the makings of the public park, this book draws parallels to persistent political and socioeconomic challenges surrounding digital leisure architectures.

For instance, if we go back to the early 19th century, we witness the birthing of a radical act across several cultures and nations: the demarcating of certain public space for primarily leisure purposes. From India to the United States, for the masses, public parks became a symbol of democracy, openness, and freedom as they emerged from a protracted struggle with the state or imperial powers. There was much euphoria about these urban commons and their unregulated and public character. The parks heralded modernity and a new age of civility. They were places where all classes of people could congregate, serving as a unique albeit temporal terrain for social equality. Yet, on further examination, we reveal a contentious process of shaping, regulating, and sustaining the public character of the urban commons.

Interestingly, the 21st century celebrates the birth of another leisure space that shares this rhetoric of being open, free, universal, non-utilitarian, and democratic: Web 2.0. This digital commons has been looked upon as a site where regardless of gender, age, and/or culture, people commune, browse aimlessly, socialize, and share their views openly. Yet, two decades later, the usage of social network sites reveal tremendous political, economic, and sociocultural tensions. Their usage opens debates of critical concern on what constitutes the common good. Governments and corporations are finding ways to control and mediate users through strategic architecting and managing of the digital commons. Meanwhile, online consumers, hackers, and activists are harnessing these sites for a range of activities and causes. This book argues that the digital commons and the urban commons are hardly dissimilar, as they both confront an uphill battle in the preservation of their public spheres.

By drawing parallels between public parks and Web 2.0 spaces, this book highlights the historicity and plurality of public leisure spaces and provides a much needed rootedness in the highly speculative media discourse. While social network sites have a relatively short history, for decades the study of underlying structures, networks, and cultures have been of core preoccupation in the fields of geography, urban planning, and sociology. The public park, be it the classic 19th century park to the more contemporary corporate and fantasy park, serves as a spatial metaphor to reveal different aspects of these new media spaces. This book makes the case that the history of developing public parks across cultures provides a rich source for understanding the political and commercial battle for public leisure topographies.

Using metaphors to understand the Internet is not new. In fact, to conceptualize the Internet, the metaphor is never far behind. To explain 'new' technology spaces and activity, there has been a need to look at the 'old': the unfamiliar turns to the familiar to make itself known. In talking about spaces of social interaction online, we find ourselves in virtual *dungeons*, *pubs*, cyber*cafés*, chatrooms, *home*pages, online *communities*, and MUD *lobbies* (Adams, 2005). In situating ourselves in larger virtual geographies, we're confronted with the electronic frontier, or caught on the information superhighway. In fact, the need to architect a sense of place online has become a paramount strategy in understanding digital social life. Thus, resorting to material space to explain the virtual realm is hardly an uncommon practice.

This book leverages on this comparative approach to cater to a largely overlooked aspect of digital space, that of leisure. The text focuses on a specific yet universal spatial construction within (and across) cities since the 1900s—the public park. The proposition argued here is that if the Internet is a 'city' as Mitchell (1996) popularly states, then its online common leisure spaces are its parks. The blurring of the virtual and the real is thereby an evolving social and spatial interaction and construction.

So, in viewing Web 2.0 through the lens of public parks, historically, transnationally, and transculturally, the intent is to reveal the complex polity in creating and sustaining such spaces. This disrupts the popular notion that leisure is largely non-contentious, with little overt economic, utilitarian, and/or 'productive' value or predetermined goals. Hence, this book investigates a range of park spaces to make transparent the diverse needs and deeds of actors in the leisure commons, both offline and online.

THE LEISURE COMMONS

One can say that we have come a long way from the Puritan perspective of leisure as sin; today, it is the prime commodity of social life (Chudacoff, 2007). Nowadays, there are faint memories of public leisure spaces as contentious. For the most part, this memory remains buried in the chronicles of a bygone era. Thereby, few question the presence of public parks. While strolling through manicured landscapes, fewer still glimpse their embedded controversies.

Meanwhile, much attention is being paid to online leisure spaces such as Facebook, Twitter, and YouTube, where people check each other out, share their views on movies, or just mindlessly browse. This is seen as the mark of the 21st century. It is the arrival of a new kind of movement, a novel means of experiencing, producing, and consuming leisure: "whether desired or not as part of any 'official' history of this currently central cultural medium, online recreation or 'virtual leisure' has been positioned among the dominant elements within the Internet's development" (Weiss, 2006, p. 961). What is more, these activities are seen as perhaps the most democratic of all. As such, these common social spaces appear to serve as open platforms for all to participate, circumventing gender, class, nationality, and culture (Arora, 2011).

In fact, the relationship between technology and leisure is highly debated. For instance, there is a belief that new communication technologies produce new kinds of leisure. Here, traditional practice gives way to novel acts of leisure. Another school of thought highlights deteriorating social ties and lifestyles through remote and isolating leisure practices online (Turkle, 2012). There is no denying that new technology terrains inspire contemporary forms of leisure expressions and enactments. Rather than focusing on the differences in leisure topographies, the goal here is to delve into the similarities between the digital and the urban commons. The intent is to demystify major claims of novelty by grounding the Web 2.0 hype in situated and historical contexts of public parks. Hence, the starting premise of this book is the following: In order to understand new digital space, we need to move away from viewing it as technical and see it more as a cultural and social space. While new technologies open up new possibilities for performing leisure, these tools are still rooted in a basic human impulse that has found expression in different ways over the centuries (Roberts, 2006).

In recent ethnographic studies of social network sites, a range of online leisure activities have surfaced, from game play to dating to just plain sociality

(Arora, 2010; Boyd, 2007; Buckingham & Willett, 2006). Researchers have found that online leisure takes on numerous and often discriminatory forms and is shaped by economic, political, and sociocultural forces. To maintain their democratic and public status, these spaces are in constant flux. Similarly, public parks in the early 19th century reveal struggles between the state and their citizens when shifting from privatized to public leisure domains (Rosenzweig & Blackmar, 1992). What constitutes as a public park has expanded to corporate, community, and fantasy parks, signifying new trends in leisure spaces. These trends resemble niche and semi-private social networking sites. Overall, the problematic that pervades the cyberleisure realm can be addressed through park comparisons, especially along the lines of the following. (1) *Open versus closed systems*: what are the social costs to keep these lived spaces open and who regulates participation? (2) *Private versus public interests*: what is the impact of commercializing and branding of leisure spaces on their diversity? And (3) *work versus play dimension*: how much labor goes into producing leisure online and how are corporations appropriating these spaces to enhance productivity?

SPATIAL, HISTORICAL, AND TRANSNATIONAL FOCUS

Initial Internet investigations were deliberately dissociative from physical place (Adams, 2005). The Internet was seen as the new egalitarian arena for the 21st century. This utopian declaration came with a dystopian reaction. Emphasis was placed on the digital divide, where more than two-thirds of the world's population are digital 'have-nots' (Arora, 2010). This 'novel' space becomes a terrain that perpetuates inequality. People often simulate behaviors from the physical world within virtual space. Today, it is commonly understood that to make sense of online space, we need to look at their offline counterpart (Baym, 2009). Indeed, it is hard to say to what extent these practices are novel and a product of the contemporary time. This book takes both a historical and spatial approach to understanding the cyberleisure terrain. It compares real and virtual leisure spaces, particularly between public parks and Web 2.0 spaces. In comparing these topographies—their histories, architectures, regulatory structures and diversity—we address issues of corporatization, privatization, and homogenization.

Granted, a Japanese garden is different from its English counterpart. Yet, this book argues that, worldwide, there are common patterns across the public leisure commons that are useful for analysis. Where the overlap ends, the novelty of the digital commons begins. This book starts with the generic image of what a park is and goes to demonstrate its pluralistic, cross-cultural, and contemporary nature, expanding the notion of 'public parks.' This is much like how Web 2.0 is perceived—social networks are generic and universal, yet are deeply diverse and niche-oriented, targeting different needs and cultures.

Most studies on Web 2.0 delve into the technological but not the spatial history of leisure practice online. Hence, this book fills the gap by bridging the urban and the digital commons. Given that most of the world's population struggles to gain access to the Internet, the choice of public parks (that is more accessible) as a comparative leisure terrain is a deliberate means of including such populations in this analysis. To investigate the spatiality of the leisure commons, the book has drawn heavily from the disciplines of urban planning, cultural geography, and law to address the contemporary concerns in new media studies.

For greater methodological efficacy in using metaphors to explain and critique, throughout this book specific online phenomena and concerns are juxtaposed against the physical spatial equivalent. This demonstrates the extent to which social practice has historical underpinnings. It also mitigates some of the hype on the novelty of these new digital geographies. This enables an extension of important discussions on the rights and responsibilities of citizens, and the role of the state and corporate entities and other interest groups in the makings of the leisure commons. As mentioned earlier, metaphors have been harnessed repeatedly to map virtual leisure spaces. Today, there is common acknowledgment that the real and the material realms are deeply intertwined and cannot be extricated from one another. There is an implicit agreement that the Internet has spatial characteristics in common with real-world places. After all, in situating these conceptions within real space, we are able to avoid "a purely technological interpretation and [recognize] the embeddedness and the variable outcomes of these technologies for different social orders" (Sassen, 2002b, p. 365).

NOVELTY AND LIMITATIONS OF THIS STUDY

While many books in new media studies have paid heed to the real–virtual blurring of boundaries, few works push this line much further. To do so, this book not only touches upon different disciplines but immerses deeply into them. As mentioned earlier, the fields of urban planning, law, and geography have come to the rescue in a significant way. These fields have been particularly confronted on the subject of borders between the digital and the urban commons. For instance, mediatization of the urbanscape via mobile technologies challenges urban planners and geographers. From the school of law emerges some of the most compelling literature on the subject. This should not be surprising as there is urgency in resolving how the digital domain and the physical domain can be legally bound by the same laws.

As expected, when we cross disciplines, we always encounter a new lens of analysis and different theoretical approaches. For instance, in Chapter 4 on walled gardens, we borrow the framing of gated communities as 'architectures of fear' and apply it to the increasing privatization of the digital leisure realm. Or we take the rich discourse on 'Disneyfication' and 'brand empires' of public space in Chapter 6 on fantasy parks and apply this to the threat of homogenization of digital leisure networks. In fact, this book is full of such borrowings. When these borrowings are resituated in the field of new media, it fosters innovative ways to look at staid media discourses. Of course, when embarking on unfamiliar disciplinary terrain, there is a tremendous challenge on what to extract as significant and how to identify key thinkers in these alien fields. By no means has this book leveraged and capitalized on other disciplines to the fullest. Yet it hopes to have at least opened the doorway to interdisciplinary work that goes beyond the superficial.

Most notably, this scholarship is a tribute to the field of geography for enabling the rootedness of technology hype through emphasis on the spatial context. In fact, this research is as much a spatial history of public parks as it is about Web 2.0. Most books out there on the history of new technology trace a path back to older technologies. However, this book moves away from the artifact and focuses on the social space that contributes to the constructing, regulating, and sustaining of leisure architectures.

Furthermore, this book's usage of the term 'leisure' to frame Web 2.0 spaces underlines the core characterization of the digital commons. In doing so, we situate it within the larger discourse of our current leisure society. This effort opens a new set of viewpoints around leisure, self-expression, popular culture, and social life. That said, the most emphasis goes into how these leisure architectures are constructed, regulated, and sustained by a host of actors; there is less focus on leisure-oriented practices. Given that the field of leisure studies is rich with such analysis, this book has refrained from pursuing these well-established areas.

There is also a deliberate effort to illustrate case studies on public parks and digital formations from places such as Saudi Arabia, China, and India. This effort is made to balance the disproportionate scholarly emphasis on the United States and Western European countries. This transnational effort has revealed that, in spite of such diversity in cultures and practices, in governments and publics, there is much more in common in the politics of the public park than one would assume. For instance, in the 19th century, the rise of the public park in China was a concerted effort by the state to signify modernity and to create a common platform for socializing the masses. This has uncanny parallels with the carving of public parks in Massachusetts, where the state intended to contain their 19th-century immigrants through the offering of a common recreational space. Likewise, we see common threads when we look at their respective digital spheres. As illustrated in-depth in the following chapters, we reveal that users play with digital space through humor and satire and foster communities online, both in China and in the United States.

However, we must admit that in spite of this overt bias to draw from studies in non-Western regions, there continues to be a strong focus on the West. Partly, this is attributable to the fact that with wealthier nations, come greater support for research. The case of China stands as an exception to the usual Western focus in new media studies. In fact, there are some excellent recent studies on the Chinese Internet. This work is driven by the urgency to understand China as a digital and urban giant of the 21st century. When it comes to research on public parks, the United Kingdom is particularly impressive: they come with a rich heritage on the urban commons, such as the Speakers' Corner, the Victoria Park, and other signature public leisure spheres. As we will see in the coming chapters, the United Kingdom was a pioneer in the making of the public park and influenced significantly the direction of the then unique and unprecedented terrain. This, coupled with strong and specialized departments in garden studies, reveals their national passion for the politics and history of gardening. Such national research pursuits has added richness to this text.

It is important to note here that this book is not exhaustive in its focus on key thematic issues. Each chapter has been concertedly written with a specific focus on a particular issue facing new media studies. This by no means suggests that it is comprehensive. For instance, the protest parks chapter focuses on digital activism and social movements online. Corporate parks focus on digital and free labor; fantasy parks on branding and homogenization of public leisure networks; global parks investigate the internationalization of hyperlinked leisure networks and the walled gardens chapter digs deeper into the central question of the right to privacy on social network sites. In the next section, a more elaborate description is provided on each of these chapters. While these are indeed important and current concerns, this book should not be seen as a handbook on new media issues and concerns. If the book were to be expanded, it would examine the rise of cultural parks and how public leisure domains, such as museums, are taking to the digital sphere. It would also examine darker aspects of leisure spaces, such as pornography and terrorism. While the issues have been selective, this book is a demonstration that we can thoughtfully examine Web 2.0 concerns by stepping away from media scholarship. We can learn from multiple disciplines, and these disciplines can significantly shape our understanding of social spaces.

Overall, this interdisciplinary, comparative, and historical approach to cyberleisure space offers opportunity to gain insight into possible futures of Web 2.0 spaces in particular and of contemporary public leisure territories in general. It is valuable to achieve a macro-perspective on current state activities as they engage and experiment with how to architect and regulate online leisure spheres. This book does not delve much into why people do what they do online. It is best to leave the 'why' to a systematic, longitudinal anthropological study. This book reveals how an effort to move beyond one's conventional discipline can lead to innovative thinking on a subject. No doubt, the comparison between public parks and social network sites has its limitations. However, by stretching this metaphorical frame, we can gauge to what extent contemporary social spaces have been reproduced from the past. Further, when there is difficulty in overlap between these topographies, it should be viewed as an opportunity to delve into the possible uniqueness and novelty of digital leisure networks. Interestingly, this book may give a new lease on life to park studies, given their diverse and pluralistic contemporary forms.

OUTLINE OF CHAPTERS

The leisure commons is made up of several types of public parks. This book allows for specific avenues of investigation to flourish by creating a typology of parks, namely, protest parks, corporate parks, walled gardens, fantasy parks, and global parks. This typology serves up appropriate metaphors to frame discussions of contemporary concern in new media studies, namely, the corporatization, privatization, and the homogenization of digital leisure networks. Below, each chapter is briefly described. This should give the reader a sense of the way in which the subject of public leisure space has been approached, bridging the real and the virtual.

In Chapter 2, 'Metaphor as Method: Conceptualizing the Internet through Spatial Metaphors,' the foundation is laid to explain the methodology of this book. The usage of spatial metaphors as cognitive devices is not new. Nor is its utilization in conceptualizing the Internet since its inception. This chapter goes in-depth about how metaphors enable us to map and visualize digital landscapes and communicate their cultures to a wide-ranging audience. This chapter describes the full spectrum of popular metaphors of the Internet, from emphasizing its borderless cartography to capturing browsing movements through its space. The Wild Wild Web, the Frontier, the Cloud, and the Electronic Ghetto are some of the metaphors that have played a part in evoking expectations and emotions as well as endorsing policy and practice. There is a section devoted to understanding the 'digital flâneur' and how this is tied to historical practices of movement within public leisure environments. We address notions of anonymity in digital wandering and commodification of such navigations. Another section reveals strong linkages between the metaphors of the urban commons, the digital commons, and the leisure commons that serve to explain the title of this book. By examining the radical history of the public commons, we borrow concepts such as the common good, the tragedy of the commons, and its reclamation in this digital and urban age. Here, we investigate what constitutes as the public good and ways we can sustain it.

In Chapter 3, 'Protest Parks: Digital Activism and the Public Leisure Sphere,' we draw parallels between the use of public leisure spaces, such as parks and squares, and the use of certain forms of digital networks for protest. Similarities between these two sorts of social contexts are worth considering, particularly in relation to their political dimensions. This effort situates the current conversation about political mobilization via social media into dialogue with the historical analysis of public parks as protest spaces. Specifically, public parks were, in a similar fashion, designed for leisure and consumption but were often appropriated as sites of resistance. It brings together literature on urban parks as centers of democracy and literature on new media spaces as portals of cyber-protest, extending the spatial history of digital politics. Here, 'protest parks' serve as a metaphor for contemporary digital networks of activism. The chapter also examines the range of mediations that enable the transformation of these seemingly innocuous spaces into places of activism. Particularly, it reveals the social architecture of and political enactments within public parks and squares in the United States, the United Kingdom, and China in conjunction with protests within their contemporary digital networks. We discover that protests do not so much detract from the park's (or Social Networking Site's, SNS) primary leisure purpose but often are deliberate products of such infrastructures. Further, depending on the regulatory mechanisms, we see protest taking on more creative, play-like forms of expression, creating new rituals of communication between citizens and the state. Finally, we see a plurality of democracies emerge through the complex interplay of the public–private nature of leisure space and political action. Overall, this chapter reveals how politics and leisure are historically and dialectically tied between the real and the virtual.

In Chapter 4, 'Walled Gardens: Online Privacy, Leisure Architectures, and Public Values,' the focus is on how web architectures are being walled in, dictated by market systems and state ideologies. These cyber-enclosures are justified along the lines of privacy that garners protection, efficiency, and functionality. There is significant concern for the potential and irrevocable loss of the 'public' and 'open' character intended for Internet infrastructures. Some fear the fostering of social segregation, homogenization, and corporatization of leisure and a loss of civic sense. This chapter addresses these concerns by looking at contemporary material architectures that are shaping public social and leisure space. Particularly examined here are gated communities, shopping malls, children's playgrounds, and guerrilla and community gardens. This chapter argues that for a comprehensive understanding on privacy and public leisure architectures, we need to recognize the parallels between these virtual and material spheres as social norms, values, and laws permeate their boundaries. Vigorous debates pervade over the governance and architecture of Web 2.0 leisure geographies, particularly in the areas of privacy, property, paternity, and profit. At the heart of these discussions is the growing concern over whether we are losing the battle in maintaining the digital public sphere as a non-commodified and unified domain. There is also concern over whether this is the fate of the leisure commons. Explored here are the persistent pursuit of idealism, public values, and human ingenuities in the transformation of fortified enclaves into inclusive and communal spaces. Basically, the metaphorical application of the 'walled garden' is used to emphasize the extent of openness and freedom in the public character of these leisure territories. Walls are built to secure and protect people online and offline and, yet, these very structures can be confining or liberating. Whether it is a physical or a virtual walled garden, we can learn much about its ideology by examining its architecture and the range of practices within such space. This comparison illustrates aspects of inclusion and exclusion, commercialism, protection, activism, and the regeneration of public leisure spaces. Here, privacy is seen through the lens of accessibility, choice, and ownership and reified through certain architectural trends of public leisure space—gardens within gated communities, urban malls, playgrounds, and community gardens.

Chapter 5, 'Corporate Parks: Usurping Leisure Terrains for Digital Labor' examines new workspaces in the physical and virtual domains and the expectations of new work cultures. There is a shift in perception of what counts as a space of productivity. This corporate usurping and appropriation of leisure spaces is becoming visible across different sectors and across the globe, manifesting in technology, industrial, science, and/ or information parks. Simultaneously, we see corporations embrace and inhabit social and leisure spaces online: think Blogger, Facebook, and Twitter. This is seen as enabling the restructuring of the private-sector model from top-down to a more employee-driven and customer-oriented culture. This chapter focuses on this new trend of corporate leisure spaces intended to foster innovation, networks, and communication in this global and social media age. It synthesizes online and offline workspaces across geographies. It addresses this new architecting of workspaces and relates it to labor, leisure, innovation, and networking in business culture. This chapter applies the metaphor of 'corporate parks' to examine how business geographies extend to and influence social media spaces as they strive to realign the labor and leisure domain for innovation and employee satisfaction. This trend is positioned historically by examining how leisure space has been legitimized over time to increase productivity. Such an examination highlights the implications of mobility in the architecting of work environments. This builds on the blurring boundaries of work and play and sheds light on contemporary social media trends, such as digital and free labor. This chapter draws from the historical struggle for leisure in the labor landscape such as hobby farming, factory gardens, and the role of social visionaries in bridging these two domains. Also, it reveals the nature of contemporary technopoles, converging ecosystems of diverse companies that foster networks and serve as corporate incubators. Lastly, it examines the current trend on work cafés and its impact on entrepreneurship.

Chapter 6, 'Fantasy Parks: Consumption of Virtual Worlds of Amusement' emphasizes the historicity of public spaces of fantasy and how they were reflective of the public values, sentiments, and social transformations of the time. By looking at the precedents of such leisure spaces, as well as contemporary manifestations, we attend to the shifts in the cultural tone of society toward the notion of fantasy. Here, 'fantasy parks' serve as a compelling metaphor for the understanding of digital amusement ecologies of virtual worlds and digital gaming platforms. This chapter investigates the complex interplay of citizens, corporations, and the state in the makings of such immersive fantasyscapes, both virtual and material. The sanitized and predictive quality of spaces such as Disneyland threatens to homogenize public leisure domains. Yet there are always localizing dynamics that stem from indigenous interpretation, play, and representation of generic icons of fantasy and Western-oriented mass media narratives. Strong emphasis is placed on the notion of Disneyfication of fantasy that has pervaded both online and offline terrain. Much of this is applied to gaming platforms, such as the World of Warcraft, and virtual worlds, such as Second Life, as well as new mobile gaming apps. This chapter reveals the brand empires that structure these landscapes and the template that is marketed across the globe.

Lastly, Chapter 7, 'Global Cities, Global Parks: Globalizing of Virtual Leisure Networks' examines the globalization and cosmopolitanism of digital and material leisure networks. It begins by making a case for a more vibrant ecology of public leisure space. To do so requires the dismantling of conventional boundaries between the park and the city. Further, this chapter uses the metaphor of 'global cities' to emphasize the hierarchies in digital leisure networks. These global cities serve as command centers and fulcrums for the industrial, the creative, the leisurely, and the privileged, as well as temporal laborers and the migrant class. Similarly, not all social networking sites share the same power and influence. For instance, Facebook and Twitter are the virtual command centers of the digital age. Hence, this chapter makes the case that global cities and digital leisure networks function similarly: both are at once stateless and yet constrained by diverse national laws and local social practices.

Overall, this book fragments the leisure commons into disparate park formations. This metaphorical cornucopia allows the reader to go on multiple journeys to investigate protest, fantasy, work and play, and the globalization of leisure and privacy.

2 Metaphor as Method
Conceptualizing the Internet through Spatial Metaphors

> We should be careful not to fall into the trap of either declaring that cyberspace provides new public spaces or that cyberspace further weakens public spaces in the geographic domain. Instead, we should seek to document the socio-spatial relations of cyberspace, the interplay between public and private concerns, and how these intersect with geographic space . . . they are spatialisations utilizing a geographic metaphor to gain tangibility.
>
> Dodge and Kitchin, *Mapping Cyberspace*

Common understanding of online space has transformed substantively since its inception, revealed, for instance, in the shift in terminology from 'cyberspace' to 'Web 2.0.' There is now an acknowledgment that virtual space is not a monolithic structure but a plurality of networks shaped by a range of stakeholders. Since the first decade of euphoria about the internet, there has been a growing demand to anchor these spaces in real-world infrastructures rather than accept the initial interpretation of such spaces as revolutionary, unprecedented, and novel (Arora, 2012b; Baym, 2009). Metaphors have been faithfully employed in this pursuit, clarifying and making tangible the unknown through the known (Lopez, 2003). There is a clear mission to architect social media spaces through experienced and experiencing physical structures such as chat*rooms*, electronic *frontiers*, *home*pages, and information *highways*. Focusing on the spatial dimension emphasizes the importance of the underlying structure and its nature and design in shaping online social action. Thereby, spatial metaphors are particularly useful and powerful instruments to foster a deeper understanding of digital space.

Such rhetoric has been harnessed strategically across disciplines: scholars of law draw upon metaphors to transfer legal code from physical to virtual worlds (e.g., Lemley, 2003); scholars of policy use metaphors to simplify and communicate technological novelty and justify new commitments from ecommerce to egovernment and elearning (e.g., Sawhney, 2007; Osenga, 2013); scholars of architecture and urban planning celebrate the metaphor as it serves as an important reminder of how central their field is in shaping new public space—their design strategies help to construct online social networks (e.g., Wilson, 2001).

While metaphors are aplenty to explain, argue, and normalize Web 2.0 spaces, they are often engaged in a peripheral manner. Scholars rarely delve deeper into how they are created, sustained, and transformed through social action. There is a need to attend to these debates and the points of departure where the metaphor fails to explain the novel phenomenon. This effort can be viewed as an opportunity to extend the conversation on relations between culture and social structure to the online sphere (Pachucki & Breiger, 2010). Further, by capturing the shift in the nature and usage of metaphors for digital space, we can get a sense of the dominant concerns of our time. Thereby, there is a need for a framework that organizes and deepens our understandings of emergent spaces of Web 2.0 by situating them in material architectures. This book seeks to meet this fundamental need.

THE VIRTUE OF THE METAPHOR

Rhetoric as a Cognitive and Policy Tool

The metaphor today enjoys central status. Over the past half century, metaphors have shifted position from being discussed peripherally to now being an essential part of conceptual reasoning (Johnson, 2010). Seen in prior days as a mere figure of speech, it is now viewed as a critical cognitive device that allows us to unpack complexity and normalize novelty by extending the meaning of content/context to which it is applied (Raffel, 2013). In other words:

> a metaphor makes sense of something by tying it to another, more familiar image. An alternative model holds that a metaphor creates an association between dissimilar things, inflecting disjunct meanings to create a new, third, meaning . . . on this account, a metaphor does not contain meaning; it provides a starting point for the construction of meaning. (Adams, 1997, p. 156)

Paul Adams, a professor of geography, compares metaphors to myths where shared worldviews are fostered through these tools (1997, p. 156). He reminds us of the classic work of Stanley Deetz, George Lakoff and Mark Johnson, who provided three kinds of metaphors to help frame social practice: *positional metaphors*, which translate non-spatial conditions into spatial orientations; *ontological metaphors*, which provide a theoretical connection between phenomena; and *structural metaphors*, which make associations among familiar life experiences. Whatever metaphorical tool we choose, when it comes to its relationship with technology, Adams underlines how it fosters lateral thinking and enables people to innovate and be creative.

The sociologist Jose Lopez (2003) argues that the usage of metaphors is unavoidable in social theory and yet continues to be neglected. Metaphors, he states, can be powerful instruments for social theorists to understand 'structure,' not constructed as opposed to or in association with agency but, rather, constitutive of each other. Being powerful, however, does not necessarily make these analogies correct. Lopez highlights some influential metaphors with strong linear and deterministic leanings that have penetrated deeply into the social sciences. For instance, the 19th-century sociologist Emile Durkheim drew parallels between society and the living organism and, with that, transferred the concept of evolution to social structure and practice. This work continues to shape theory today. Karl Marx proposed the 'base-superstructure' metaphor to frame relations of production, fostering generations to view power relations through this causal lens. And the sociologist Talcott Parson utilized the language of thermodynamics to explain social action, dangerously taking us on the mechanistic path of viewing social domains as systems of equilibrium, homeostasis, and adaptation. These examples no doubt underline how significant and potent metaphors can be in our understandings of social life.

What is particularly useful for this book is the 'architectural metaphor' approach. We can capitalize on this approach to reveal networks of concepts that serve to map social reality. This can have a cognitive impact, especially when it is generative and not just descriptive of a particular social domain. Using the example of the 'eye is a camera' metaphor, Lopez illustrates how common knowledge of the camera can serve to explain the complexities of the eye and produce new vocabularies to capture this novel phenomenon. However, he warns us of the danger of metaphors where, instead of being transformative, they can serve as mere transfer mechanisms of meaning:

> A transfer also sets up a relationship between the host domain and another phenomenological domain; however what distinguishes it from a transformation is that it fails to produce new domain specific concepts, meanings, and theoretical strategies in the host domain. (2003, p. 16)

When applying metaphors to abstract domains, it is important to remember that we do not entirely transfer all meanings from one context to another. Instead, the aspect that is in most need of comprehension is tied to the original and familiar context.

A metaphor is not just a cognitive tool to map and comprehend social reality and aid in social science research. A metaphor can also be a policy tool to communicate and convince vast audiences of new initiatives. Based on how it frames an issue, it can push policy agendas in wide-ranging fields, such as immigration, telecommunications, education, and war (Lakoff

& Johnson, 1980; Lemley, 2003; Sawhney, 2007). An example that has gained attention (and is described in more detail in the following chapters) is the usage of the 'information highway' metaphor to describe the access and speed of connectivity of the Internet. Mark Stefik (1996), a principal scientist at the Xerox Palo Alto Research Center (PARC), argues that this early metaphorical adoption severely limited public perception and expectation of the digital domain. Its use narrowed conversations to the issue of access of information via digital networks, neglecting more complex matters on information usage. Interestingly, Stefik is currently leading projects in the domain of knowledge, language, and interactions at PARC to pioneer ways in which to deal with issues of congestion, public safety, and public policy. To capture such complexities of urban life, he frames these projects through the metaphor of 'social nervous systems.' Hence, metaphors can be as much a cause of policy as they are a product of such enactments.

More recent metaphors of the digital domain such as the 'cloud' have influenced our practices and information policy perspective in a compelling way. Kristen Osenga, a professor of law, traces the long-standing influence of metaphors in policy-making and reasons for their popular appeal as well as their downfall. The shift from framing the Internet as a 'highway' to that of the 'cloud' allows for a change in focus: the Internet is no longer a conduit but rather an amalgamator of content (Osenga, 2013). Media coverage on the cloud gives the impression that content is 'out there' in a nebulous form with no tangible ownership. This coverage gives the misleading impression that surveillance is less likely. This metaphor has, however, caught on, and it is now a common phenomenon that resonates with the general audience. Rather than a scientific angle, there is more of a romantic and poetic leaning in interpreting the cloud, encouraging the mystique of the Internet rather than making it more concise. Thereby, we must remember that metaphors are not neutral but come with a historical sociolinguistic usage that influences popular perception and informs policy-making.

Mapping Real and Virtual Geographies

Mapping is a powerful way to geographically visualize digital landscapes and communicate their cultures to a wide-ranging audience. Maps as graphic tools classify, represent, and communicate spatial relations within the database of information that we inhabit and co-construct. In fact, as the Internet becomes more ubiquitous and dominant in our daily lives, there is a need to use all possible tools of communication for engagement and entertainment. There is much truth to the old adage of a picture speaks a thousand words: a single visualization of data can shift public perception more easily than extensive survey records. We need to keep in mind that maps do not represent scientific facts or an all-encompassing reality. Instead, they are social constructions that emphasize certain digital cultures for the lay public. For instance, given the historical roots of the Internet stemming from the Cold War crisis, this virtual geospace can be perceived as more of an Americanized spatialization, especially in the initial years (Dodge & Kitchin, 2000).

Take a more current example of Twitter and the possible implications in the strategic mapping of the geography of this digital platform (Graham, Stephens & Hale, 2013). After collecting a significant database of tweets, a graphic and a metaphor were coined to illustrate such data: the 'spatially aware treemap.' Its purpose was to highlight the proportion of a country's Twitter representation and to underline the empowering effect of the specific geography of Twitter. Most studies emphasize the tremendous inequality in the geography of content through distinct divisions of labor, where the global North prevails as the predominant producer. However, through this 'treemap' metaphor, it was revealed that out of the six largest information producers through Twitter, namely, (1) the United States, (2) Brazil, (3) Indonesia, (4) the United Kingdom, (5) Mexico, and (6) Malaysia, only two of the countries came from the global North. Also, when examining these cartographies, we need to be aware of the purpose of their production and the contextual nature of their messages and embedded values.

While no doubt our interest in mapping has piqued substantively with the creation of the Internet, historically, thinking spatially has been the cornerstone of geographic thought. Michael Curry (2005), a professor of geography, argues that there are four main notions of space used in mapping concepts: Aristotelian, whereby space is static, hierarchical, and concrete; Newtonian, whereby space is a kind of absolute grid within which objects are located and events occur; Leibnizian, whereby space is fundamentally relational and defined entirely in terms of those relationships; and Kantian, whereby space is conceptualized as a form imposed on the world by humans. Hence, when critically examining the orientation of mapping metaphors for the Internet, it may serve as useful to identify the spatial framing's philosophical leaning.

Popular metaphors of Internet mapping cover a wide spectrum, from constructing a borderless cartography to deeply architecting these digital realms as extensions of 'real' infrastructures. Some scholars emphasize the notion of the borderless digital realm. In the early years, Novak (1991) framed the Internet as a 'liquid architecture' of data flows impervious to boundaries while Benedikt (1991) compared it to a 'common mental geography.' Howard Rheingold (2000) popularized the collapsing of boundaries when he proposed early on that the virtual sphere offered opportunities to meld *gemeinschaft* (where community relationships are locally bounded and tied to social status) and *gesellschaft* (where community relationships are individualistic, impersonal, and private). Such metaphors were instrumental in pushing the agenda of the globalizing and converging of digital cultures and mind-sets. Martin Dodge and Rob Kitchen's classic book *Mapping Cyberspace* (2000) contradicts the popular rhetoric of the 'death of distance,' where it was believed the digital sphere would collapse all sense of borders. Instead, they argue that space became even more important in reconfiguring and expanding our notions of social practice, online and offline. In other words, geography continues to matter, be it virtual or real.

A way of bridging the real and virtual mapping of metaphors can be seen in Nina Wakeford's (1999) classic ethnography of the first of the London cybercafés. She proposed that we view material-technical networks as the intersection of three landscapes of computing: the *online landscape* concerns the visual and textual practices on the machines; the *expert landscape* involves the actors, such as the engineers, technicians, and others who contribute to the workings of the machine; and the *translation landscape* alludes to the intermediaries who serve as interpreters for customers and staff, such as the 'cyberhosts.'

It is understandable how at the birth of the Internet, there was a dominant trend to disengage and disassociate from physical place. What was most alluring about the new digital sphere was that it promised a new public sphere where "ideas of citizenship, commonality, and things not private, but accessible and observable by all" could come to fruition (Papacharissi, 2002, p. 9). Viewing contemporary public space as limited and limiting, to even a failed arena, virtual space became associated with individual self-fulfillment and personal development. A new democratic dream was born. Part of this disassociation came with the denial of form to this nebulous structure. By disregarding conventional mappings onto digital space, it was seen as being liberated from the shackles of real-world boundaries and territories. Taken further, it was compellingly argued that chronic power inequities are embedded in our physical world and unlike this reality, the virtual sphere is inherently free.

However, with every utopian declaration comes a dystopian reaction. Universal access and usage of the Internet continues to be unrealized. There are tremendous disparities in access: while the United States enjoys almost 80 percent access, only 15 percent of the African population has the same opportunity (Internet World Statistics, 2012). Thereby, this 'novel' space becomes yet another realm for manifesting and perpetuating inequality. Turning the tables, euphoria is replaced with apprehension—there is a fear of this 'elite space' becoming the new "playground for the privileged" (Hess, 1995, p. 116). This time, however, it is at an unprecedented and globally ambitious scale that threatens to exponentially increase the divides between people and cultures. Such utopia–dystopia visions and proclamations confront histories and current practice wherein people often times evoke, imitate, and simulate behaviors and spaces of the physical world within virtual spaces.

In other words, the novelty of the technological realm does not automatically determine novelty in social action. So, for instance, virtual communities, some argue, are not being formed as enthusiastically as expected with inherently new and noble social rules. Rather, they are often building on and

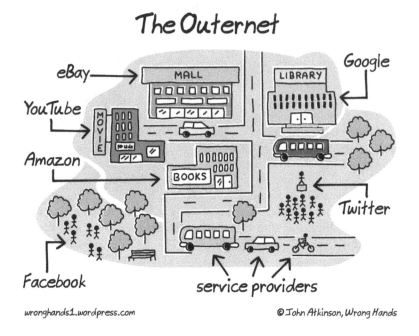

Figure 2.1   The Outernet: Facebook and the park metaphor.
Source: Cartoon © John Atkinson, Wrong Hands, used with permission.

extending offline relationships. When faced with such a dichotomy, we can't help but pay respect to the rich conceptual framing offered by Michel Foucault (1967), who offers us a way out of this constrained environment. His 'heterotopia' term, a third place that has "the curious property of being in relation with all . . . other sites" powerfully turns our attention away from the staid interplay between the ideal and the real, between utopia and dystopia.

Worth noting is that the educational, political, religious, and other spatial practices are seen as being fundamentally integrated with the online realm (Koehler et al., 2011; Leighley, 2010; Lundby, 2011). Blurring of these online and offline spheres has inspired new avenues of scholarship. Such scholarship embarks on fleshing out the embedding and embodied aspects of this new perspective. Indeed, we have moved significantly from the initial exceptionalist claim, at least in theory. However, when it comes to regulating such spaces, we keep slipping back into these conceptual traps, viewing digital space as virtual 'property,' distinct and demarcated from real place (Cohen, 2007). What is needed is more of what Lawrence Lessig (1999) terms as 'structural plasticity,' where we understand that much of online and offline regulation is contingent on ordinary and informal social norms, like peer pressure and reward systems. At the same time, we need to acknowledge the underlying codes of regulation that influence digital behavior. These codes are much like the rules of conduct influencing public enactments in parks, squares, streets, and other material contexts. From isolated and exclusive worlds, the virtual and the real are now enmeshed and entangled; a cornucopia of realities now inhabits this united space. Hence, our focus on metaphor usage is not just to highlight and comprehend the novel aspects of these new media spaces but also to connect these disparate user-generated online spaces to a more coherent and multiplexed model.

SPATIAL METAPHORS OF THE INTERNET

Wild Wild Web to the Electronic Ghetto

It is no coincidence that one of the most popular initial spatial metaphors employed to grasp the Internet domain was that of 'cyberspace,' albeit not without controversy:

> Adherents of the 'cyberspace' metaphor have been insufficiently sensitive to the ways in which theories of cyberspace as space themselves function as acts of social construction. Specifically, the leading theories all have deployed the metaphoric construct of cyberspace to situate cyberspace, explicitly or implicitly, as separate space. This denies all of the ways in which cyberspace operates as both extension and evolution of everyday spatial practice. (Cohen, 2007, p. 210)

As previously stated, in digital space's first years there was a deep bias for exceptionalism. It was a time of utopic expectations, a hope for the digital Shangri-la. The Internet metaphors of the 1980s and 1990s reflected as much, exciting audiences with the possibility of having new identities, new terrains, and new beginnings for a democratic life. This feeling of novelty and unchartered territory was captured through other popular metaphors of that time, namely, the 'new frontier' (electronic frontier) and the 'Wild Wild West' (Wild Wild Web). The Internet was paralleled to the 'Western Frontier,' where "land was free for the taking, where explorers could roam, and communities could form with their own rules. It was an endless expanse of space: open, free, and replete with possibility" (Hunter, 2003, p. 442–443). This metaphor carried with it the notion of the American 'land of opportunity' and the 'pioneering spirit.' Such a comparison moved beyond the mere geophysical domain to that which is ideological—a space of limitless possibility and individual agency. This resulted in the popular conviction that the Internet needed to be kept free from state regulation to maximize individual potential.

The Electronic Frontier Foundation (EFF), an organization created in 1990 to preserve freedom of speech and equality in access to digital networks, adopted this metaphor wholeheartedly. John Perry Barlow, a founder of the EFF, released a manifesto outlining the organization's vision for the Internet, with a particular goal of keeping the government's hands off this new space.

> Over the last 50 years, the people of the developed world have begun to cross into a landscape unlike any which humanity has experienced before. It is a region without physical shape or form. It exists, like a standing wave, in the vast web of our electronic communications systems. It consists of electron states, microwaves, magnetic fields, light pulses and thought itself . . . Governments of the Industrial World, you weary giants of flesh and steel, I come from Cyberspace, the new home of Mind. On behalf of the future, I ask you of the past to leave us alone. You are not welcome among us. You have no sovereignty where we gather. (Barlow, 1996)

By emphasizing the unprecedented nature of this space, Barlow justified the stance that outside forces (particularly the United States government) had no right to regulate this new 'landscape.' This document became deeply influential and arguably contributed to the stance the government took in those years, leaving this space relatively open and unregulated. Instead, the government placed their faith in the common man's ingenuity and creativity within this digital geography.

Thus the metaphor of the 'frontier' has been applied to push for deregulation. Yet it is crucial to gain a more balanced perspective. If we are to genuinely gain insight into the nature of the Internet, we need to keep in mind the trade-offs that ensued during this supposed golden era of the Western frontier. Alfred Yen, a professor of law, beseeches us to pay careful attention to the spectrum of debates surrounding metaphorical comparisons. By capitalizing on the complexities of the past, he argues that we can enhance our understandings of the present and create insightful planning for future policy, especially considering how often rhetoric gets reified. In comparing the Internet to the Western frontier, Yen remarks that contrary to being a limitless land of opportunity for all, the frontier was actually a domain where deep injustice pervaded. He argues that historical episodes of exploitation during that time could have been prevented through legal regulation. If anything, this metaphor served as propaganda for "the supposed 'American' character traits as inquisitiveness, inventiveness, practicality, independence, diligence, restlessness, and exuberance" (2002, p. 1211). Besides, there is deep-seated romanticism surrounding the metaphor of 'wilderness' for the Internet, without acknowledging serious security issues. The price for digital freedom can be very high: "This seeming absence makes cyberspace a dangerous wilderness characterized by free pornography, spam, identity theft, rampant copyright infringement, gambling and hacking" (p. 1223).

Another conceptual problem with this metaphor was that it implied an 'empty' space that needed to be filled, waiting for structures to be built by the users themselves. This was misleading: In actuality, from the start, users craved and even demanded predictable and secure architectures within which they could conduct online dealings and build trust online (Cohen, 2007). In fact, many viewed this space as a marketplace and hence pushed for legal protections to be extended to this online sphere. Today these concerns are primal and have gained urgency as we discover ourselves being watched by the state. What we understood to be deeply private affairs online have become commercial databanks available to the highest corporate bidder. In fact, today there is demand for thoughtful Internet regulation framed with the human rights discourse along with an urgent need to reexamine our metaphorical conception of digital space.

Another influential Internet metaphor of this time was the 'information highway,' driven by its utilitarian ethos. In the 1990s, then United States Senator Al Gore popularized this metaphor as a way of underlining the importance of the ambitious undertaking of high-speed communication systems (Benjamin & Wigand, 1995). A global information infrastructure was being formulated, with numerous alliances across states and the private sector. To converge different technologies such as the telephone, television, and the computer, there was a need to partner with several and often competing organizations. Connectivity was and arguably still is the byword. Hence, it was essential to provide reassuring communication to the lay public to demonstrate that this ambitious interconnectivity project was necessary and not alien:

> A network of highways, much like the interstates of the 1950s . . . highways carrying information rather than people or goods . . . and it's not just one eight-lane turnpike, but a collection of interstates and feeder roads made of different materials in the same way that highways are concrete or macadam or gravel. Some highways will be made of fiber optics, others of coaxial cable, others will be wireless (Gore, 1993, December 21; as cited in Adams, 1997).

Further, this was reinforced at the U.S. Senate commerce committee where Senator Hollings advocated this vision of the Information Age:

> Simply put, fiber to the home, school and business is an essential infrastructure for economic development in the Information Age of the 21st century, just as railroads were in the last century, and *highways* [italics added] were in this century (Hollings, 1990, September 12; as cited in Benjamin & Wigand, 1995).

This metaphor is no doubt useful to communicate the scaling of technological infrastructures. Yet, if propelled further, it can garner insight into the actual implementation of such endeavors. For instance, Sawhney (1992) capitalized on the different stages in the development of highways to structure and predict the growth pattern of the telecommunications infrastructure: "the generalizability of this model is tested by comparing its conceptual framework with the historical data on the development of the highways and automobiles" (p. 541). Through this metaphorical strategy, he was able to trace and apply recurring patterns of how a new technology space starts as a feeder to the established system and then goes onto displacing that system with the development of long-distance capabilities.

By paying attention to the socioeconomic and political issues encountered in the scaling of highway infrastructures, we can perhaps foresee some of the contemporary social concerns when scaling digital infrastructures. For instance, can we readily assume that the information superhighway is to be uncritically celebrated? Can we take on the assumption that highways are inherently and universally good and have been instrumental in connecting disparate communities, opening spaces for a more egalitarian flow of goods, jobs, and services? Will this infrastructure benefit one and all as promised regardless of economic and/or social status? What exactly is being connected across sectors? What is really moving— ideas, goods, or services? Who are the direct and indirect beneficiaries? Hence, if we are to apply metaphors to novel phenomena, we should also see this as an opportunity to probe deeper. We should allow the learned wisdom of the familiar terrain to guide us in asking relevant and essential questions of unfamiliar architectures.

In his book *Internet Dreams: Archetypes, Myths, and Metaphors* (1996), Mark Stefik suggested an alternative to the 'information highway' metaphor. He was one of the earlier scholars who proposed the digital realm as multiple spatial archetypes, emphasizing the utility aspect of information instead of its access. Stefik suggested the adoption of four metaphors, each focusing on a different facet: Internet as the (1) digital library, (2) electronic mail, (3) electronic marketplace, and (4) digital world. Basically, he drew parallels of each of these metaphors to the Jungian archetypes, the keeper of knowledge, the communicator, the trader/warrior, and the adventurer, respectively. Dominant in this model is its utilitarian nature, which focuses on the usage of information and places less emphasis on the cultural dimensions of social media space.

Of course, when we talk about the possible crumbling of boundaries and barriers, we can't help but tie globalization into this conversation. Marshall McLuhan's (1962) renowned metaphor of the 'global village' gained a new lease on life with the onset of the Internet. There continues to be a faithful group of media ecologists that parade this metaphor and give it robustness in this Web 2.0 era. That said, the attempt to move away from McLuhan's implicit Western orientation led to the popularity of Arjun Appadurai's (1996) 'technoscapes' metaphor. It contested the Americanized perception of the digital sphere (made obvious with the adoption of the

'Western frontier' metaphor) and put emphasis on connectivity and mobility within and across diverse cultural spaces. His metaphor shifted focus away from the homogenization of online space. Instead, it offered a new conceptual framework to grasp the globalizing nature of the virtual sphere. The emphasis lay on the flow between virtual spaces to highlight contemporary global configurations between nations, technologies, people, and ideas. However, the trade-off here is that it sacrifices the architecting and grounding of digital spaces and pushes it into a nebulous space that appears unbounded and seamless. Interestingly, in his recently released book, *The Future as Cultural Fact* (2013), Appadurai eloquently addresses the need for spatial rootedness by privileging the historical to gain insight into current social architectures.

> We need to recognize that histories produce geographies and not vice versa. We must get away from the notion that there is some kind of spatial landscape against which time writes its story. Instead, it is historical agents, institutions, actors and power that make the geography. Of course, there are commercial geographies, geographies of nations, geographies of religion, ecological geographies, any number of geographies, but each of them is historically produced. They did not pre-exist so that people could act in or with them. To perceive histories as producing geographies, promises a better grasp of the knowledge produced in the humanities, the social sciences, and even the natural sciences about the way in which regions, areas, and even civilizations emerge from the work of human beings. (p. 66)

In other words, if we are to truly understand the extent to which these spaces are novel and unprecedented, the historical is a capable and essential architect of Internet space.

Not all metaphors of that time were wide-ranging and ambitious as the global village or technoscapes. While in the initial years the Wild Wild West, the Western frontier, and the information highway were dominant Internet metaphors, other metaphors have sprung up based on specific contexts and causes. There is a logic pervading the architecting of digital space: in linking virtual contexts to specific places, we set the scene for understanding human communication. Take, for instance, the metaphor of the 'lobby' that was applied to the American Online (AOL) computer-network service. By using this specific context, it propelled an investigation of the temporal nature of situated practice. To put it simply, people are often passing through versus 'residing' in digital space. This metaphor also succeeds in emphasizing the high affordance of serendipity where people bump into each other online, get acquainted, and chat, much like the organic workings of movement within a lobby. Another similar metaphor that sought to highlight a social aspect of digital space is the 'virtual pub.' In her book, *Hanging Out in the Virtual Pub: Masculinities and Relationships Online*

(2002), Lori Kendall succeeded in evoking and transferring the personality of place to online communities. Through a rich ethnography of BlueSky, a specific online group, she captured the range of conversation that occurred in that space from the trivial and humorous to the more rowdy and argumentative, much like a typical pub.

While certain places like pubs and lobbies do not come with major baggage, others can be highly provocative and confrontational. Such places serve to inspire waves of emotion, from euphoria to wrath. The 'Electronic Agora' is one such metaphor, coined to emphasize the perennial yearning for a genuine public sphere. This draws from the Aristotelian concept of the 'agora,' the urban center and heart of public life and the birthplace of democratic governance. The Internet was expected to rise to the task and at last provide a forum accessible to all. In *City of Bits: Space, Place, and the Infobahn*, William Mitchell (1996) poetically captured such naïveté of the time:

> So the Net eliminates a traditional dimension of civic legibility. In the standard sort of spatial city, where you are frequently tells who you are. (And who you are will often determine where you are allowed to be.) Geography is destiny; it constructs representations of crisp and often brutal clarity. You may come from the right side of the tracks or the wrong side, from Beverly Hills, Chinatown, East Los, or Watts, from the Loop, the North Side, or the South Side, from Beacon Hill, the North End, Cambridge, Somerville, or Roxbury—and everybody knows how to read this code. (If you are homeless, of course, you are nobody.) You may find yourself situated in gendered space or ungendered, domains of the powerful or margins of the powerless; there are financial districts for the pinstripe set, pretentious yuppie watering holes, places where you need a jacket and tie, golf clubs where you won't see any Jews or blacks, shopping malls, combat zones, student dives, teenage hangouts, gay bars, redneck bars, biker bars, skid rows, and death rows. But the Net's despatialization of interaction destroys the geocode's key. (p. 10)

Not all metaphors of that time reflect such naïve or, to be kinder, optimistic perspectives. In fact, a number of studies came out in critique of these utopic notions of the Internet, emphasizing instead virtual exclusion, colonization, and segregation within the infosphere (Castronova, 2001; Gunkel & Gunkel, 1997; Sardar & Ravetz, 1996). The 'electronic ghettos' metaphor emerged to emphasize the confinement and entrapment of the poor and marginalized. The 'information black holes' captured the trappings that gender, race, ethnicity, and class play in limiting access and opportunities within this supposed open digital realm (Graham & Marvin, 2005). Barriers of poverty, illiteracy, impairment of vision, and learning disabilities are held responsible for confining certain disadvantaged populations to their own socially deprived realm, online and offline. And while these populations are seen to continue their lives in electronic ghettos, they also coexist with the promised lands of the Internet that are well beyond reach. Hence, contrary to democratic and utopic notions of the Internet, these perspectives reveal a different reality with deep segmentations, segregations, and social struggles.

Given our experienced inhabitation of digital space, one would imagine that today there are more critical metaphors pervading our discourse on the Internet. However, utopic notions continue to persist: astonishingly, Gunkel and Gunkel (2009) find that two decades later, these metaphors have found their way into architecting and designing massively multiplayer online role-playing games (MMORPGs) with little acknowledgment of past critiques. Such peripheral engagements with rhetoric happen time and again, perpetuating misleading conceptions of the past and flawed rationales for the future. Hence, when transforming virtual space through the metaphor of real space, it is essential to situate our understandings in existing and historical, social and cultural practices. Baym (2009) argues that paying heed to sociocultural behaviors and relations reveals diverse and contesting practices and a more nuanced understanding of the Internet as a plurality of cultural spaces.

For the most part, contemporary scholars agree that there is no uniform, universal, and monolithic digital sphere but a host of niche spaces within this domain. Also, there is some agreement on the fact that the Internet is not a novel and utopic space but a realm shaped by sociocultural action and human relations. Today, the usage of the 'Web 2.0' and 'Network' metaphors emphasize the Internet's hyperlinked navigational affordances as well as its deep and persistent sociality. Given the dominance of the search engine and social networking sites in our day-to-day lives, the web pays heed to the complexities of movement, pathways, and ways of navigating. The next section will illustrate this point further.

Analogies of the Online Navigator

User movement influences digital space and vice versa. The inhabitant's activity gives life to a place. In fact, all geographies require enculturation through the quotidian act of the user, converting it to a meaningful social space. If we neglect to pay attention to the navigational flows of a user, we would be condemning our perspective to a static construction. In pioneering the actor-network theory, Bruno Latour emphasized the need to focus on the performance of individuals within a network to gain insight into the makings of structure. He lamented that oftentimes scholars use the term 'network' in a staid fashion, giving insufficient weight to user activities, movements, and navigations. Hence, Latour remarks that 'networks' have come to mean "transport without deformation, and instantaneous, unmediated access to every piece of information." (Latour, 1999, p. 15).

At the initial stage of the Internet, studies demonstrated that users exercise cognitive mapping strategies to navigate their virtual environments. They do so in a way that is similar to how they approach real spaces (Richardson, Montello & Hegarty, 1999). For example, an image of the user was perceived as a lone ranger, the web 'surfer' riding the virtual waves that came her way. The user's digital 'trail' was recorded as she clicked on certain links and visited certain pages, influencing the pathways on which the system was being shaped. Not surprisingly, the 'Footprint' system, an online monitoring system that collates user behavior and movement, appears to be metaphorically synced. The choice of trails and waves evokes a sense of the user being part of the 'wilderness,' a popular metaphor addressed in the prior section. Consistent with the 'frontier' metaphor, users were seen as 'explorers' and 'pioneers,' expected to create 'settlements' through their laborious activity.

Over time, this tamed the wild landscape. Soon after, the Internet began to be framed by the metaphor of the 'city' as popularized by Mitchell (1996).

> This will be a city unrooted to any definite spot on the surface of the earth, shaped by connectivity and bandwidth constraints rather than by accessibility and land values, largely asynchornous in its operation, and inhabited by disembodied and fragmented subjects who exist as collections of aliases and agents. Its places will be constructed virtually by software instead of physically from stones and timbers, and they will be connected by logical linkages rather than by doors, passageways, and streets. (p. 24)

The urbanization of digital space inspired new analogies of movement: highways and intersections. The user's navigation generated web 'traffic,' bringing attention to certain sites over others. Access and speed were now privileged in discussions on Internet movement. Classic texts such as Michel de Certeau's *The Practice of Everyday Life* (1984) were drawn upon to eloquently illustrate the relations of spatiality and movement in the 'walking through the city' metaphor. This metaphor was most influential in negotiating the 'strategies' of mapping used by institutions, corporations, and the state, as well as that of the individual's play with urban space. The visualization of the city's plan was influenced by the 'tactics' at the street level, where everyday moves through these power grids resist, reshape, and reconstitute the purpose of these mapped spaces.

In addition, a city cannot exist without a marketplace or a 'bazaar,' seen as the functional heart of urbanity. Of course, the privatizing and commercializing of what was imagined to be a utopic and democratic public sphere or 'digital agora' did not transpire without resistance. Eric Raymond's famous article, *The Cathedral and the Bazaar* (1999), raised the alarm for the critical need to keep these spaces open to the public. This work inspired the free and open software movement that is alive and well today.

The movement seeks to create alternative landscapes that embed a shared and communal ideology. In fact, a rich hacker culture thrives today, where several thousand developers work tirelessly to sustain this ethos. They are proud parents of the open software Linux, a powerful manifestation of subversive space online, contesting the notion of 'private property' within the digital domain. Of course this process is messy given that developers from different nations and with diverse agendas come together to structure a place that is meant to be holistic.

> The Linux community seems to resemble a great babbling bazaar of differing agendas and approaches . . . out of which a coherent and stable system seemingly emerges only by a succession of miracles. (Raymond, 1999, p. 24)

Yet miracles do happen and are happening time and again with the rise of the Wikipublics and the Creative Commons and Anonymous. Here, the user is portrayed as a co-architect of digital space, an active participant rather than a passive beneficiary of the digital world.

Lest we get swept away by this romanticism, we need to remember that contestation happens both ways; consumption is a powerful elixir in the virtual terrain. As social networking sites become more commercialized, they often resemble malls more than public squares. With a tremendous variety of online sites to experience and navigate, the user can be viewed as an 'online tourist' for peripheral engagement and entertainment. She strolls through the web, consuming space in a distant manner. Eliciting sentiment and emotion becomes one way in which to sustain the virtual presence of the transient electronic masses. That being said, how much of this strolling is driven purely by the interests, intent, and innate desires of the user?

In this age of personalization, there is much discussion on how information comes to the user versus how they seek it. The metaphor of the 'filter bubble' captures the fear that our browsing activity is not free as we believe (Pariser, 2011). Instead, it is mediated by powerful algorithms that reflect the interests of corporate giants who would like to entrap users within bubbles of homogeneity and predictability. This is perhaps a call for help: We do not want to lose the spirit of the 'digital wanderer,' to cease to explore unchartered territory, and to expand the frontier of our minds. Hence, it is worth investigating the nature of our navigations through spatial metaphors to detect the extent to which our actions are still spontaneous and free. Spatial metaphors for navigation range from scanning (covering a large area without depth), browsing (following a path until a goal is achieved), searching (explicit goal search), exploring (finding the extent of information), and wandering (unstructured search) (Canter, Rivers & Storrs, 1985).

For those with a more adventurous streak, part of the allure of being a 'digital wanderer' is getting lost and navigating our way back. In Nicholas Burbules's essay 'The Web as a Rhetorical Place' (2002), he points out that the web we encounter and the web we make are not just a matter of convenience but can be a tremendous source of learning. He connects mobility—which he sees as fundamental to navigation—to that of knowledge building. Hence, software affordances and constraints serve as learning challenges: users learn to play with their privacy settings on browsers and social networking sites, and learn to protest and access pirated media material through creative means.

While the digital wanderer is not linked explicitly to the domain of commerce, the 'digital flâneur' has been popularly used to capture the commodification of this wandering, which is delved deeper into in Chapter 6 on fantasy parks. Inspired by Charles Baudelaire's early modernism figure, the 'flâneur,' the user today, is seen as walking around these digital spaces as a detached observer. The metaphor of the 'digital flâneur' emphasizes the unstructured, detached, and playful movement of the user that is believed to pervade while web browsing. Functioning as an urban tourist of sorts, the user's primal goal is to absorb as much experience as possible rather than faithfully entrench in one place. Through this analogical framing, one's location becomes a concern. Another key element emphasized through the digital flâneur metaphor is the promise of anonymity.

> The crowd is his element, as the air is that of birds and water of fishes. His passion and his profession are to become one flesh with the crowd. For the perfect flâneur, for the passionate spectator, it is an immense joy to set up house in the heart of the multitude, amid the ebb and flow of movement, in the midst of the fugitive and the infinite. To be away from home and yet to feel oneself everywhere at home; to see the world, to be at the centre of the world, and yet to remain hidden from the world—impartial natures which the tongue can but clumsily define. The spectator is a prince who everywhere rejoices in his incognito. (Baudelaire, 1964, p. 9)

The impression of the user is that she is unseen in this digitally mediated environment. She can, from the comfort of her bedroom, voyeuristically gaze at the world outside through the homecam. At once, the digital space is intimate and yet distant. She is thereby exempt from typical social rules of engagement and interaction. In the early history of the Internet, this perception of user anonymity gained popularity through an unlikely candidate: a cartoon in the *New Yorker* on July 5, 1993. The cartoon by Peter Steiner portrayed two dogs, one sitting on a chair in front of a computer, the other sitting on the floor. The dog in the chair tells the other dog, "On the Internet, nobody knows you're a dog." Today, this dialogue has gained much complexity as we are now watched just as much as we continue to watch others. For example, we live in the time of PRISM, the recently revealed scandal of the United States government surveillance system that aggregates data on all citizens' mundane activities through portals such as

Skype, Facebook, Twitter, and Google. In such a time we can hardly profess to be truly anonymous. Hence, from the nascent stage of the Internet as 'wilderness,' where regulation was looked upon as suppressive, today there is a growing demand across the globe to architect and institute the protection of our rights. Such action is necessary if we are to prevent electronic ghettos, filter bubbles, and digital malls from completely taking over digital space. In other words, we need to create constraints on digital terrains if we are to navigate more freely, more anonymously, and more playfully.

Urban Commons, Digital Commons, and the Leisure Commons

The Internet did not pan out as planned. One may say it is human nature to take control, to contain the wilderness, and to erect gates around the frontiers. We may believe that in the age of the metropolis, online or offline private property is essential and inevitable. Trust is for the naïve. Perhaps, the agora was part of Greek wishful thinking, evoked time and again to perpetuate more fantasy. Maybe it is natural for us to be dreamers, to help cope with reality. But then, as touched upon earlier, what are we to make of the rise of the Wikipublics and the Creative Commons and Anonymous? Are they just a small band of idealists amid a larger sphere of realists? To put this in perspective, if we are to look at the top-ranking sites in the world that are most visited and inhabited in 2012, unsurprisingly we find Facebook, Google, YouTube, Yahoo, Amazon, and Baidu at the frontlines (Alexa Rankings, 2012). The empires of commerce are well entrenched and, indeed, the Internet does resemble a marketplace. But move one step further and we encounter Wikipedia, ranked fifth on the global index for the most visited sites.

Unlike other websites with clear business models based on advertising revenue and data mining, Wikipedia is a digital commons that survives (and thrives) based on mass donations and volunteerism. Jimmy Wales, Wikipedia's founder, repeatedly calls out to the user to financially support the organization so it can continue to provide free and public service to all. In 2011, his letter went public on the Wikipedia website:

> Google might have close to a million servers. Yahoo has something like 13,000 staff. We have 679 servers and 95 staff. Wikipedia is the #5 site on the web and serves 450 million different people every month—with billions of page views.
>
> Commerce is fine. Advertising is not evil. But it doesn't belong here. Not in Wikipedia. Wikipedia is something special.
>
> It is like a library or a *public park* [italics added]. It is like a temple for the mind. It is a place we can all go to think, to learn, to share our knowledge with others.
>
> Jimmy Wales
> Wikipedia Founder

The metaphorical comparison with the public park to capture the character of the digital commons is not new; it comes from a radical history of the public commons. In *Capitalism 3.0: A Guide to Reclaiming the Commons* (2006), Peter Barnes diverts attention to the heritage of the commons. He argues that we are slowly losing the commons unless we become active in sustaining it. The commons traditionally refers to the natural environment that is shared by all. As societies became more urbanized in the 18th and 19th centuries, the commons came to be associated with a deliberate carving of nature for mass consumption—the public park. It served an important social function as a shared space for fostering communities and networks in a dense urbanity. The fundamental nature of the commons is that it resists commodification and ownership: it belongs jointly to the members of the community. In return, members are responsible for preserving and sustaining this space for the future generations.

It is natural to be skeptical about this arrangement, as equality in participation seems to be an ideal. Surely some members are more abusive than others, and some are more generous with their time and effort to preserve this public commons? This concern was popularized by Garrett Hardin's seminal article, 'The Tragedy of the Commons' (1968). The usage of this metaphor illustrates how an individual's interest conflicts with the common interest. In this article, Hardin provokes with a question: When the resources of the commons is overused and depleted by overuse, who is accountable? Given that some members are more likely to use these resources than others, are all of them equally responsible for the sustenance of this shared domain? Given that human action is driven by self-interest, it is seen as impossible to sustain this unique public property regime that rests on a shared belief and understanding. This essay was deeply influential for opponents of the communal ownership of resources. While this logic seems compelling, the reality is that urban parks are found across the globe and new social norms create a shared understanding of expectations in this public sphere.

In fact, several scholars across disciplines rose to defend the urban commons against Hardin's neoliberal worldview. One of the most famous critiques stems from Elinor Ostrom (Ostrom et al., 1999), the Nobel Prize–winning economist as she embarks on re-examining Hardin's assumptions on joint governance. She argues that contrary to Hardin's belief that depletion of the urban commons is inevitable because of mass usage, she believes that it is in people's self-interest to cooperate to sustain shared terrains. Users *do* come together to collectively discover ways to restrain themselves and perpetuate and enhance these common grounds. They do so particularly when they perceive this as having high social value. Besides, we cannot ignore the fact that all law relies on the shared belief of the citizens. If anything, the urban commons is proof that the human act is not a simple product of rational choice, but is part of a wider social network of norms and values.

No wonder the 'urban commons' serve as a powerful metaphor for the digital commons, defined as:

> informational resources created and shared within voluntary communities of varying size and interests. These resources are typically held de facto as communal, rather than private or public (i.e. state) property. Management of the resource is characteristically oriented towards use within the community, rather than exchange in the market. As a result, separation between producers and consumers is minimal in the digital commons. (Stalder, 2010, p. 313)

Scholars have appropriated this metaphor to delve deeper into the makings of the Internet and its extent of spatial freedom. This has not just been applied to the general Internet realm, but also appropriated to the cultural space online, framed as the 'cultural commons.' Here, the principle of sharing one's creativity, ideas, and inventions on common digital platforms like YouTube, Flickr, and Instagram captures the growing spirit of Web 2.0. Such a spirit can be viewed as user-generated content for all to consume. This has also inspired the metaphor of the 'creative commons' that shields these collective creative products from the powerful forces of patenting, proprietary control, and market triumphalism. These forces seek to envelop culture with commercialism. Dan Hunter (2003) termed the steady encroachment of digital space by private actors as the tragedy of the digital 'anticommons.' In this scenario, multiple parties can effectively control the digital resource and prevent others from using it. Therefore, the space is of use to nobody.

Hence, in this book we investigate the fundamental tension between the rational and the ideal, the public and the private, and what is expected of human action in the shaping of their common space. The 'leisure commons' metaphor articulates the core dimension of urban parks: As explained in the introduction, these parks are marked for primarily non-instrumental social usage by the public. While indeed each chapter faithfully follows a specific typology of urban parks—protest parks, fantasy parks, walled gardens, global parks, and corporate parks—to illustrate dimensions of corporatization, commercialism, and privatization, they are part of the larger rubric of the leisure commons. This stands as a constant reminder that the digital commons has its history in the urban commons, and that leisure is a key instrument defying the utility logic within this shared geography.

3   Protest Parks
Digital Activism and the Public Leisure Sphere

From public parks to allotments, squatted community gardens to the 'polemic landscapes' of peace or fascist gardens, as well as the 'defiant garden,' the plot is the territory under discussion, the patch of earth where it all happens.

George McKay, *Radical Gardening: Politics, Idealism & Rebellion in the Garden*

Parks, squares and other shared spaces have been seen as symbols of collective well-being and possibility, expressions of achievement and aspiration by urban leaders and visionaries, sites of public encounter and formation of civic culture, and significant spaces of political deliberation and agonistic struggle.

Ash Amin, 'Collective Culture and Urban Public Space'

As a new infrastructure of public spaces, the Internet promises to restore some of the critical functions of the public sphere. It does so through the unique combination of communication media and public space, for the Internet is both a medium and a space. It is global in both these properties. Like the bourgeois public sphere, Internet technology itself was the product of modern capitalism and nation-state. And like that historical public sphere, it seems to encroach upon the power of both, this time by loosening territorial and human barriers. Whether a re-feudalization of online spaces will occur depends on the result of future struggles, including the struggles to resist the commercialization of the Internet and to maintain a high level of engagement in public debate, civic association and protest on the Internet.

Guobin Yang, *The Internet and the Rise of a Transnational Chinese Cultural Sphere*

Public protest takes on many forms. Picture bongo drummers and weed, tucked away in a corner of San Francisco's Golden Gate Park. Passersby today can still see the remnants of the hippie counterculture movement of the 1960s, a time of Allen Ginsberg and his poetic activism, a time of 'human be-ins,' live music, Frisbees, and flying coffee can lids protesting the Vietnam War. The cultural studies theorist Stuart Hall marked this as an 'American moment'—a subculture of peace, love, community, and self that successfully fostered the symbolism of a Peace Park. In early 1967, 20,000 hippies gathered in Golden Gate Park to protest a California law that made LSD illegal. True to the Summer of Love ethos, this public expression in a park brought together artists and the Hell's Angels and took on the guise of chanting mantras and singing Grateful Dead tunes. And every now and then, these lived politics of space get reignited and revived to serve a contemporary cause. In September 2012, a group in San Francisco calling themselves Space TranSFormers took to the park to demonstrate against the growing commercialism of the city's public spaces. The organizer Ryan Rising told the *San Francisco Examiner*:

> This is a big tradition in San Francisco—people gathering freely to share music and discussions . . .We feel like it's kind of fading away . . . We will be thinking about how to grow our own food, heal each other with herbal medicines, build natural structures . . . this is about reaching a permanent relationship of balance with the earth.

Across the globe, it is possible to witness temporal inhabitations of public parks and squares. These represent mass expressions of social discontent and focal points of resistance. Tahrir Square in Cairo, not coincidentally known as 'Liberation Square,' gained international attention when 50,000 protesters gathered there during the 2011 Egyptian revolution. It was a revolution that brought about the downfall of the then president, Hosni Mubarak. That same year people congregated and camped in Zuccotti Park in New York to voice their concerns and anger against the economy, the government, and a range of social issues. In fact, the choice of urban parks and squares for public protest comes from a deep tradition of mass political activism spanning across nations (Arora, 2011; Forthcoming; Mitchell, 1995; Williams, 2006).

This is not to say that urban park spaces are exclusive sites for mass activism. Undoubtedly, unrest can be found on the streets and beyond. But there is a difference between streets and other public spaces that are continuously usurped for mass protest. Streets can take on a temporal meaning of social solidarity, but urban parks were concertedly designed for democracy, making this investigation particularly noteworthy. If we are to look at the historical emergence of urban parks from the 1800s onward, their spatial design and diverse inhabitation, it is astonishing to learn how embedded political action has been within these public domains.

Events such as the Twitter revolution have ignited passions and expectations in the virtual realm. Some see these technologies as the new public sphere of mass activism. In this sphere, governments, corporations, and

Figure 3.1   Planting democracy in Tahrir Square.
Source: Carlos Latuff/Wikimedia Commons/Public Domain.

citizens compete to hijack these platforms for their own agendas. This is an age where many express euphoria about the rising use of online network spaces as novel sites for political mobilization and expression.

A journey down memory lane can serve as a humble reminder of the rootedness of these now virtual actions in public parks. The geographer Ash Amin underlines that historically these spaces were "key sites of cultural formation and popular political practice. What went on in them—and how they were structured—shaped civic conduct and politics in general" (2008, p. 5). However, he does alert us to our sentimentality in expecting today's urban public parks to hold the same influence they had on civic culture and political formation as in the past.

By drawing parallels between the historic use of public parks and squares in the city, and the use of certain forms of digital networks such as Twitter and Facebook, we can gain a more integrated and critical understanding of the novelty of these spaces. Similarities and differences between these two contexts are worth considering. This is particularly true given their political dimensions: Such a comparison puts the current conversation about the use of Social Networking Sites (SNSs) as tools of political mobilization into dialogue with the historical analysis of public parks. In other words, public parks and SNSs can be seen as spaces that, in a similar fashion, were designed for leisure and consumption but are also appropriated as sites of resistance.

To some extent, the protest parks of the past can serve as a metaphor for contemporary digital networks of sociality, leisure, and activism.

Comparing these parks to digital networks can better explain the relationship between virtual and material public space and their role in political movements. In the early years of the Internet, the historian Mark Poster (1996) lamented that new media space threatened traditional public spaces of political action:

> The issue of the public sphere is at the heart of any reconceptualization of democracy. Contemporary social relations seem to be devoid of a basic level of interactive practice which, in the past, was the matrix of democratizing politics: loci such as the agora, the New England town hall, the village Church, the coffee house, the tavern, the public square, a convenient barn, a union hall, a park, a factory lunchroom, and even a street corner. Many of these places remain but no longer serve as organizing centers for political discussion and action. It appears that the media, especially television but also other forms of electronic communication isolate citizens from one another and substitute themselves for older spaces of politics. (p. 207)

Additionally, Dawn Nunziato (2005) later argued that digital space was being usurped by the private sector and was, in fact, more controlled than the material public sphere. She believed the material public sphere still enjoyed a certain amount of public freedom:

> The absence of public forums in cyberspace augurs the absence of meaningful protection for free speech under the First Amendment. In real space, the existence of government-owned property as a forum for speech available to all comers provides an important guarantee for such speech. In contrast, as increasingly more speech takes place in cyberspace, the affirmative constitutional protections for free speech that exist in real space are in danger of being sacrificed. (p. 1118)

Today the conversation grows in complexity. Perspectives are not about substitution but about the unique amalgamation of the digital and the physical realm in enacting public protest and the regulation that pervades across these boundaries.

The main focus of this chapter is to examine why certain public leisure spaces attract political action while others do not. In order to understand why digital leisure platforms serve as protest spaces today, we benefit from a historical approach by examining urban parks and squares and how they transformed into spaces of protest. Hence, protest parks serve as a metaphor to remind us of the spatial history of political and social activism in public leisure domains.

The chapter also examines the range of mediations that enable the transformation of these seemingly innocuous spaces into places of activism. Here, the argument is illustrated by comparing the social architecture of and political enactments within urban parks and squares in the United States, the United Kingdom, and China with cyber-protests within their contemporary digital networks. It becomes clear how material and virtual leisure platforms have evoked similar reactions: Some are enthusiastic about these platforms, which are seen as a significant expansion of democracy into public space. Others, however, take a more dismal view of the platforms as prime spaces to disarm and manipulate the masses through their seemingly unregulated leisure character.

In analyzing events and movements that started within urban park locales across these nations, this chapter reveals how politics and leisure are historically and dialectically tied. In focusing on the range of social movements across park and digital geographies, we discover that protests do not so much detract from the park's primary leisure purpose but often are deliberate products of such infrastructures. Further, depending on the regulatory mechanisms of these urban parks, we see protest taking on more creative, play-like forms of expression, creating new rituals of communication between citizens and the state. Finally, we see a plurality of democracies emerge through the complex interplay of the public-private nature of leisure space and political action.

Overall, we consider how virtual leisure territories serve as centers of democracy and sites of protest. As D'Arcus (2006, p. 7) argues, "given the centrality of public spaces to political protest—and, in the media age, of the more abstract space of a mediated public sphere—careful analysis of how they come to be, how they are regulated, and the precise nature of their connection to power and dissent is essential."

DIGITAL LEISURE NETWORKS AS THE 'NEW' POLITICAL SPHERE

"Twitter has been criticized as a time-waster—a way for people to inform their friends about the minutiae of their lives, 140 characters at a time. But in the past month, 140 characters were enough to shine a light on Iranian oppression and elevate Twitter to the level of change agent." Mark Pfeifle (2009), a former national security adviser, made this argument and called for Twitter to be nominated for the Nobel Peace Prize for its role in supporting political uprisings against despotic rulers across the world.

He is hardly alone. Some argue that SNS platforms such as Facebook lend themselves to political communication far more than traditional media spaces. This is attributable to their unique design affordances (Neumayer & Raffl, 2008). As such, it is argued that their social architecture allows for groups to form more easily and information to disseminate quicker through their interactive channels. Social technology spaces here are seen as relatively decentralized 'leaderless' networks or what Coopman (2011) terms as 'dissentworks,' given their unique technological affordances for loosely distributed political networks. This parallels the new progressive political mappings where conformity and ideology have been replaced by subjectivity and diversity in activism. In fact, della Porta (2005) proposes to look at these new forms of political organization as 'relaxed framing,' which enables people to situate multiple and diverse issues and concerns within the same protest event and space. For instance, in November 2011, Marc Smith and his team from the Social Media Research Foundation in Belmont, California, mapped the political networks during the Occupy movement and found that conversations within the Twitterverse were highly diffused and segmented across diverse groups. Several clusters were identified by an algorithm that looked for 'islands' within this digital geography of protest and found a loosely-knit crowd that affiliated with a spectrum of causes under the larger umbrella of #occupywallstreet.

These 'islands' are also viewed as more socially encompassing as they are seen to create a common ethos that transcends national and cultural borders. In recent years, we have witnessed new communication platforms facilitating and coordinating protests around issues such as the environment, economy, and nuclear disarmament, blurring what constitutes a local or global cause. However, this does not indicate that these virtual protest sites have gone rogue, with complete autonomy from state authority (as argued more in-depth in Chapter 7 on global parks). Despite these bold crossings, the state continues to be a key player in mediating the movements of people and defining their rights and benefits online and offline. In other words, the cyber-protest park is an amalgamation of connectivity that is inclusive yet fragmentary, creating multiple 'glocal' spatial alliances: "the netizen might be the formative figure in a new kind of political relation, one that shares allegiance to the nation with allegiance to the Net and to planetary political spaces it inaugurates" (Poster, 2006, p. 78).

One persistent challenge has been to understand the relationship between varied protest groups and the underlying infrastructures that shape their communication. Several scholars have taken on the task of figuring out the nebulous 'mob' that repeatedly unifies for a common cause. The mob has been frequently associated with the Shakespearean canon of an undisciplined and unruly assembly of citizens driven by mere emotion. Web 2.0's participatory culture has given new impetus to reframe this mob, from a 'mindless' public to what Howard Rheingold defines as 'smartmobs.' Here, we see a dramatic shift in perception on what constitutes an intelligent network.

Extending this discussion to the political sphere, there is tremendous expectation that new social software enables and empowers groups to aggregate, connect, and express themselves cohesively on a shared concern. Examples abound on how social networks enable protest, even in the most restrictive of confines. In January 2013, a woman at a mall in Dammam, Saudi Arabia, tweeted a complaint against the religious police known as the 'Agents of the Committee for the Promotion of Virtue and Prevention of Vice' (CPVPV). She observed the police shutting down an educational exhibit featuring plaster models of dinosaurs by turning off the lights and ordering everyone out, alarming both the children and their parents:

Within minutes of the incident, a freshly minted Arabic Twitter hashtag, #Dammam-Hayaa-Closes-Dinosaur-Show, was generating scores of theories about their motives. Perhaps, suggested one, there was a danger that citizens might start worshipping dinosaur statues instead of God. Maybe it was just a temporary measure, said another, until the Hayaa can separate male and female dinosaurs and put them in separate rooms. Surely, declared a third, one of the lady dinosaurs had been caught in public without a male guardian. A fourth announced an all-points police alert for Barney the Dinosaur, while another suggested it was too early to judge until it was clear what the dinosaurs were wearing. Not a few tweets cast the incident in political terms. "Why close the show?" asked one. "It's not as if we don't see dinosaurs in newspapers and on TV and in the government every day." "They should go after the dinosaurs who sit on chairs," suggested another, seconded by a tweet who advised that dinosaurs in gilt-trimmed cloaks, the garment of choice for senior sheikhs, would make a better target. ('Forced into Extinction,' 2013)

This temporary unity across a diverse public demonstrates a vibrant collectivity, and many attribute this unity to the innate affordances of new media platforms. Here, crowds are given a more positive connotation, seen as a "multitude" that "cannot be reduced to sameness, a difference that remains different . . . the plural singularities of the multitude thus stand in contrast to the undifferentiated unity of people" (Hardt & Negri, 2004, p. 99). Virtual public leisure spaces such as Twitter are seen to give rise to a new political sphere as well as a more sophisticated and complex community of practice.

Perhaps one of the most visible proponents of social media as a political platform is the author of *Here Comes Everybody*, Clay Shirky (2008). Shirky espouses that this networked generation has more opportunity than ever before to engage in public speech and undertake collective action. While being careful to claim preordained outcomes of liberation and freedom from these architectures, Shirky does state that they "have become coordinating tools for nearly all of the world's political movements, just as most of the world's authoritarian governments (and, alarmingly, an increasing number of democratic ones) are trying to limit access to it" (2011, p. 2). Much in line with the German philosopher Jürgen Habermas (1962/1989), Shirky focuses on the underlying structures of Web 2.0 that allow for engaged dialogue among citizens, believing that in the long run they serve to expand the boundaries of the public sphere. However, Shirky does point out that while undoubtedly these social media sites are used far more for leisure and social purposes than mass political activism, they are still formidable spaces. In fact, some scholars have remarked that such leisure properties protect these sites from state censorship. A few years ago, the Internet activist Ethan Zuckerman facetiously proposed the 'Cute Cat

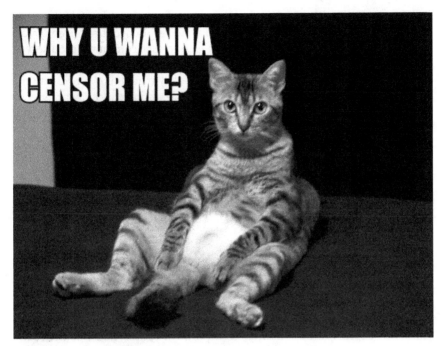

Figure 3.2   Lolcat doesn't want to be censored.
Source: By derivative work: Mr.Z-man (talk) Micas.jpg: Carlos Botelho (Micas.jpg)
[CC-BY-SA-3.0–2.5–2.0–1.0 (http://creativecommons.org/licenses/by-sa/3.0), GFDL
(http://www.gnu.org/copyleft/fdl.html), or CC-BY-SA-3.0 (http://creativecommons.
org/licenses/by-sa/3.0/)], via Wikimedia Commons.

Theory of Digital Activism' to explain the power behind banal activities
like the sharing of 'lolcat' (cute cat) videos. He argued that the entrench-
ment of online sites such as Facebook, Flickr, Blogger, and Twitter within
leisure activities makes it harder for authorities to crack down and block
them. Part of this has to do with the fact that most people are not social
activists and use these spaces for purely recreational purposes; shutting
down these leisure sites would create a public outcry.

To complicate matters, these platforms can just as well serve government
interests; thereby this places them in a legitimate 'conservative dilemma'
where the tension lies between using leisure platforms for state propaganda
versus censoring these spaces because of their potential for dissidence. We
must also not forget the economics driving these leisure infrastructures that
even the state benefits from, which poses an additional obstacle to the ban-
ning of such sites. Activists have been quick to seize this unique opportunity. After all, while Web 2.0 "was designed for mundane uses, it can
be extremely powerful in the hands of digital activists, especially those in
environments where free speech is limited" (Zuckerman, 2008, p. 1).

It is worth pointing out that already within the short Web 2.0 history, the initial hype around new media platforms is being grounded through empirical evidence. For instance, it was found that face-to-face communication served as a key factor in organizing the Egyptian revolution while only 13 percent of the protesters claimed that Twitter was the key medium to coordinate the protests (Wilson & Dunn, 2011). There is also an increasing awareness of corporate usurping of these political tendencies for profit motives: "My fear is that the hype about a Twitter/Facebook/YouTube revolution performs two functions: first, it depoliticizes our understanding of the conflicts; second, it whitewashes the role of capitalism in suppressing democracy" (Mejias, 2010, p. 4). Such corporate branding of mass activism by Twitter and Facebook are seen as a common ploy to capitalize on human struggle, turning a potential virtual public sphere into another branded empire.

In the early stages of this hype, Nicholas Garnham was quick to observe that the increasing commercialism of the Internet in general, and social networking sites such as Facebook, MySpace, YouTube, or Flickr in particular, was creating a new kind of dominance, a "colonization of the public sphere by market forces" (2000, p. 41). Malcolm Gladwell (2010) took this skepticism further and examined what constitutes as activism in the digital age. After all, he wonders, how can the mere pressing of the 'like' button on Facebook express one's solidarity to a political cause? Can this act really create an activist? He sees this superficial and peripheral engagement as 'slacktivism,' highlighting the negative effect such digital sites can have on civic responsibility and political participation. He laments, "where activists were once defined by their causes, they are now defined by their tools."

It is a good reminder to be wary of elevating every act of political affiliation and expression to that of meaningful activism. Also, it is worth keeping in mind that by emphasizing the role of the spatial and the technical in political mobilization, there is a danger of undermining the essence of mass protest—that being the deep and long-lasting sociocultural engagements of a diverse public that is struggling to be heard. Hence, while cyberprotest "that reflects the role of alternative online media, online protests, and online protest communication in society" (Fuchs, 2006, p. 275) is here to stay, it is essential to gain a more rooted and broader perspective of these platforms as domains for democracy. Therefore, this chapter examines claims of novelty and episodes of political action within virtual leisure platforms by drawing on the rich discourses surrounding a similar public leisure space—the urban park. Through this juxtaposition, we can better understand how to make sense of the hybrid positioning of these sites as propaganda, commercial, and activist spaces; how the economics of leisure space can exploit as well as protect; and how the 'relaxed framing' of such sites create an inclusive public space for political mobilization and expression. In essence, this chapter explicitly maps the relationship between public leisure space and politics, not as a digital invention but as an extension of the rich tradition of the protest parks of the past.

COMPARING 'NEW' AND 'OLD' PUBLIC LEISURE TERRITORIES

This section explores the political and historical dimension of urban parks and squares in the United States, China, and the United Kingdom as well as the contemporary usage of social media spaces within these contexts for cyber-protest. It examines the permeation of ideology across their digital and material leisure spaces and their role in the shaping of these architectures. Furthermore, it highlights a range of playful communicative modes that demonstrate the agency and creativity of the masses in harnessing these spaces for resistance.

An Ideological and Symbolic Landscape

The urban park is a narrative of spatial democracy and expressed ideology:

> From public park to garden city, there have been important moments when the garden in its most civic and municipal manifestations has been used by social movements as the site of struggle, opposition, and innovation. Sometimes, it has been the very topic itself of those activities. These moments can be short-term, temporary, crisis-ridden (as in the aggressive riot in the park), or long term and intended as permanent (as in the construction of the new green settlement). What is striking is the frequent idealism or utopianism experienced or expressed in this kind of urban public green, as though in some ways the garden itself can function as a special zone for the common articulation of social change, social experimentation, the critical rejection of some aspects of society, and even the confrontation with authority. (McKay, 2011, p. 12)

These seemingly innocuous public greens tell a story of political communication and activism, at times exhibiting the tension between authority and the masses, and between the elite and the proletarian (Mitchell, 1995). The beauty of the social engineering of public space is that intent and outcome are often misaligned as human ingenuity pushes the boundaries of these spatial imaginaries into realms that are unexpected and challenging. Intrinsic to these public leisure spaces is the fact that, across nations, they serve as a critical forum for mass dissent, capitalizing on their hybrid identity and unregulated status, at times transforming these sites into a genuine political space for the people (Williams, 2006). Oftentimes, these park spaces were instruments of the state to control and mediate the public through propaganda and were used by the private industry to seduce the consumption class. This section expounds on this above proposition, making transparent the parallel to social network sites such as Twitter, Facebook, and Weibo (China's Twitter). We will highlight how similar discussion abounds on the dictates and permeation of ideologies within these leisure spaces and the intersection of state governance, mass activism, and commerce reflecting the public–private nature of leisure space.

Let us take the example of the American urban park, designed to be a 'space of refuge' (1997 [1875]) by Frederick Law Olmsted, the famous park designer of the 19th century. Commissioned to architect parks, Olmsted collaborated with geologists, sanitary engineers, public health doctors, and social theorists to create civilized, peaceful sanctuaries where people could find refuge from the sights and sounds of the nineteenth-century city. However, this period also saw "the emergence of a symbolic landscape of protest, which often coexisted uncomfortably as a place of tourism" (Gough, 2000, p. 213). For instance, People's Park in Berkeley, California, came to be known as a public protest space with opposed, and perhaps irreconcilable, ideological visions dictating the nature and purpose of its leisure space:

> Activists and the homeless people who used the Park promoted a vision of a space marked by free interaction and the absence of coercion by powerful institutions. For them, public space was an unconstrained space within which political movements can organize and expand into wider arenas . . . The vision of representatives of the University (not to mention planners in many cities) was quite different. Theirs was one of open space for recreation and entertainment, subject to usage by an appropriate public that is allowed in. Public space thus constituted a controlled and orderly retreat where a properly behaved public might experience the spectacle of the city. (Mitchell, 1995, p. 109)

Mitchell talks about these contradictory visions: the first vision was construed as a public space where political actors shaped the functioning and scope of activity for mass protest, and the second vision was a modern conceptualization where civility, commerce, and class were privileged over what was seen as unsolicited political activity not desired by local businesses or the state. In fact, People's Park was also the spatial territory of the hippies who championed a social revolution during the 1960s. It was spaces such as this that was usurped by radicals from the Bay Area to pioneer the political outlook and cultural style of the New Left movement, launching into campaigns against militarism, racism, sexual discrimination, homophobia, mindless consumerism, and pollution. This is where the 'Californian Ideology' was born, seeping into the broader culture and influencing the values that helped shape Silicon Valley (not coincidentally the home to several new media founders of sites such as Facebook, Twitter, and Foursquare) as we know of today:

> Who would have predicted that, in less than 30 years after the battle for People's Park, squares and hippies would together create the Californian Ideology? Who would have thought that such a contradictory mix of technological determinism and libertarian individualism would

become the hybrid orthodoxy of the information age? (Barbrook & Cameron, 1996, p. 56)

Howard Rheingold (2000), the guru of this movement, advocates the counterculture values to shape the development of new information technologies and draws a vision where community activists should replace corporate capitalism and big government with a high-tech 'gift economy' or free labor for the common good. He sees this manifested through public bulletin board systems, free chat facilities, and open source software, all colluding to keep the struggle for social liberation visible and alive. As Web 2.0 geographies become more privatized, commercialized, and cordoned off by the state (as Chapter 4 on walled gardens illustrates), new media platforms continue to evoke hope in empowering the individual, enhancing personal freedom, and in radically reducing the power of the nation-state. While this may be so, it is hard to ignore that these new media platforms would not have been possible without significant infusion of capital from the American defense budget and close alliance with the corporate giant IBM. Ironically, then, efforts of users driven by the Californian Ideology to keep these spaces democratic and accessible are in some sense aligned with their corporate and state nemesis. We can go further by recognizing that this open source ideology is useful for corporations like Apple and Microsoft: The free labor of well-meaning individuals—acting as a social collective—contributes to the shaping of corporate commercial products and services. Thus, free labor also advances corporate information architectures and designs.

Of course, urban parks and their potential for mass political mobilization are not purely a Western phenomenon. Take, for instance, the Beijing Park in China during the early 20th century. On one hand, this park provided an arena for the city people to participate in China's political transformation from an imperial state to a nation-state (Shi, 1998). The Beijing Park was designed and positioned at the city center to serve as a standing symbol of social change. This stemmed from a vision of reform-minded officials who sought to transform Beijing into a 'modern' social sphere. The government intentionally designed its urban park to serve its reformist agenda of socializing the public to be modern and cultured citizens by offering free exhibitions, reading rooms, and pavilions. Such efforts emphasized the educational function of public parks. On the other hand, these spaces also served as propaganda platforms where campaigns were launched to promote public health, encourage moral behavior, and combat illiteracy. To the surprise of the government and far from intended design, the Beijing Park was used by the people for a range of purposes, at times undermining the established institutions and norms. Particularly, it served as a political forum for the "dissemination of ideas and the mobilization of the urban populace" (Shi, 1998, p. 243). It also provided a venue for pro-democracy movements and mass rallies:

Chinese parks became a public space highly contested by both the government and the civil society. On one hand, parks provided an arena for the city people to participate in modern China's political transformation. Unheard of in imperial times, frequent mass rallies held in the newly created public spaces heightened city people's demand for a political voice in national policymaking and demonstrated their strong commitment to the idea of democracy in a sovereign republic. On the other hand, the government also used the newly created spaces to push for their reformist agenda. (p. 250)

The power of the public park or square is hardly a historical anecdote as, time and again, we witness the resurgence of this leisure domain, at times with devastating consequences. The infamous Tiananmen Square stands as a sober reminder of the impact of these spaces to provoke and challenge the traditional political order. The 1989 mass-scale massacre of protesters by the state continues to stand as a testament to this political potency. This square was intentionally designed by the state as a symbol of the political might of the Chinese party. Its architecture reflected its ethos: trees lined the east and west edges of the square, framing the vastness of its center. The square was lit with large lampposts, which were fitted with video cameras, serving as a strategic space for surveillance of public leisure (Davis et al., 1995).

Interestingly, the Chinese government has approached the digital network space in a similar manner. Instead of blocking citizen Internet use, it signals its modern image on the global stage by fostering a significant and impressive digital infrastructure with the 'great firewall' surrounding its terrain (Jiang & Xu, 2009). China now boasts the world's largest Internet population of 253 million, about 19 percent of its citizenry. They have created egovernment portals across provinces to serve as local venues for citizen involvement through online chat forums. These spaces serve as symbolic architectures of democracy and progressiveness of the state authority in the information age. Research shows that some underprivileged individuals are able to publish their grievances on government websites and "even though only 7.7% of China's 137 million Internet users visit government web sites regularly, they can be a critical mass for political activism" (Jiang & Xu, 2009, p. 176). Empirical evidence of these portals revealed that in general, interactive features were not well implemented: close to half of these sites did not contain the promised chat forums. Yet places such as Zhejiang have attracted as many as 200 postings every month since its inception in 2004 and Guangdong has had a monthly average of about 1,000 entries since 2003, compared to an international average of 10 entries per month on similar government forums. It is argued that "these online structures help deter business or government misconducts and are likely to improve government image and local politics" (Jiang & Xu, 2009, p. 189).

Furthermore, China has the largest community of (micro) bloggers in the world, and they have been instrumental in exposing official and corporate misdeeds. There are now more than 200 million blogs in China. These blogs serve various purposes, from exposing corruption of local officials to contributing to the shaping of foreign policy (Hassid, 2012). The extent of the normalizing of this platform is well revealed with the rise of a new kind of celebrity *gongzhi*, or public intellectual in China—the blogger. Take the story of Li Chengpeng, a sports journalist turned activist after joining rescue efforts in the aftermath of the Wenchuan earthquake in 2008. Stationing himself from Sina Weibo, China's most popular micro-blogging platform, he fiercely criticized the poor construction of school buildings that allegedly caused the deaths of thousands of children. Since then, he has gained a loyal following of seven million who digitally gather and tune into his political commentary. The launch of his new book in January 2013, *The Whole World Knows*, a collection of his social commentaries also titled *SmILENCE*, has been blocked by the government and the readings canceled in several locations because of its political provocations. The author has been compelled to wear a stab-proof vest because of the numerous threats he has received online.

It is essential to note that much like parks that are for the most part designed for leisure, the Chinese blogosphere is also largely apolitical. Bloggers engage in social pursuits such as discussing sports, cars, the arts, and romance. Yet certain events at certain points in time temporarily transform these platforms into dynamic arenas of resistance: Political conversation seeps through and interweaves with the leisure fabric. Oftentimes censorship authorities are aware of these transgressions and yet, as MacKinnon (2008) argues, blogs "serve as a 'safety valve' by allowing enough room for a sufficiently wide range of subjects that people can let off steam about government corruption or incompetence . . . before considering taking their gripes to the streets" (p. 33). That said, there is an understanding that while these leisure sites are often deliberately designed and deployed to control and regulate dissent, they can also serve to gradually infuse the state with democratic practice. This can potentially broaden the public sphere within China as we witness time and again with the pushing of political boundaries through concerted and organized action.

Indeed, the ideology shaping these spaces can be wide-ranging, from libertarian with a strong drive towards social participation, as in the case of California, to benign authoritarianism and state paternalism, as in the case of China. However, one must not neglect the power of class over the functioning and usage of these public leisure realms. A case in point are the London parks that were strategically designed as symbols of a new capitalistic society, intent on creating a new civic and social sphere conducive for consumption (Roberts, 2001). Again, the historicity of these spaces reveals that while structured to symbolize cosmopolitanism, they instead became suffused with mass activism and public protest. As early as 1872, the British royalty understood the need for a safety valve for the masses and instituted the allowance of a 'Speakers' Corner' in Hyde Park where the public was free to express themselves. It is worth noting that at this time, Hyde Park still carried a strong connotation of being a royal park that came with a distinctive character and was a symbolic area for elite patronage. For instance, during the outbreak of the civil war, this park was repeatedly closed to the public and used by the Royalists for military purposes. Its royal lineage persisted as it continued to represent a 'pleasure garden' for the genteel. Being one of the first capitalist cities in the United Kingdom, it was designed to feed the sensibilities of a class system into the frenzied imagination of the lower strata. In other words, these civil public spheres were shaped as "a controlled environment of winding paths and closed cropped flower borders that fulfilled bourgeois fantasies of a tamed natural landscape" (Taylor, 1995, p. 386). These sites were meant to create aspiration among the working class and propel them toward the ideals of a cosmopolitan system of mobility.

In the mid-19th century, a combination of park bylaws and the use of the London County Council's venue licensing powers made the Speakers' Corner one of the few places where socialists and activists could freely meet and debate. The passing of the Lord Grosvenor's Sunday Trading Bill forbade all Sunday trading in London, including the simple act of selling fish and meat and newspapers within these public spaces. This act had the unexpected consequence of a mass protest movement, unintentionally transforming Hyde Park into a more political proletarian public sphere:

> The themes and generic styles of public meetings brought to life by Chartism now became inscribed at Hyde Park. These themes and genres translated Hyde Park into a political public sphere. Diverse and heterogeneous elements of working class interests, interests that were brought together through Chartist public spheres, were momentarily captured at Hyde Park in 1855. Several important points need to be made here. The appropriation of space by the 1855 demonstration and the genre of public speaking was therefore different in tone and substance to previous demonstrations at Hyde Park. First, the Chartist leader James Bligh took the chair and was immediately informed by Inspector Banks of the Metropolitan Police that the park was the private property of the Crown and meetings could not be held. Bligh defended his right to speak on grounds that Hyde Park was in fact public property. The Chartist leader, James Finlen, rushed to a nearby tree and immediately established another speaking space. (Roberts, 2001, p. 322)

Protesters used this social cleavage to challenge authority through public speech while still functioning within the permissibility of the system. The masses were able to appropriate and play with the space, expanding the symbolism of the urban park and thereby it's functioning over time:

> The sign 'speech' had carved out a distinct geographical and moral space in Hyde Park over a century before 1872. Constituted through

the 'last dying speeches' of the criminal class of 18th-century London, this subaltern rationality rendered visible the class character of law by disrupting the distancing of legal discourse from governance. Secondly, by undertaking a genealogical investigation of the sign 'speech' at Hyde Park, the traces left by scaffold culture were re-combined to slowly translate 'last dying speeches' into a more overtly political proletarian public sphere. (Roberts, 2001, p. 322)

To this day, London parks host Speakers' Corners where a range of social issues is covered. These Speakers' Corners reveal the fragmented and pluralistic nature of protest, less political in the conventional sense and more based in personalized and issue-based politics. Over the years, London parks have witnessed marches for disability rights, anti-austerity gatherings, antipope rallies, and cabbies against blocked lanes during the Olympics. The Speakers' Corner has become an institutionalized entity, forming a web presence and digitally consolidating around a range of projects and themes.

Fascinatingly, it now serves as a powerful metaphor for free speech, not just within London but also across international contexts. Bassem Youssef, the much loved television Egyptian presenter and political satirist, was recently arrested on the grounds of igniting public opinion by making fun of the President Mohamed Morsi. His arrest drew much media coverage, with several Arab activists defending freedom of speech in the media by evoking the Speakers' Corner to emphasize how essential it is to be able to critique authority as a sign of true democracy—"in central London's Hyde Park, people can criticize the Queen with no fear" (Alrumaihi, 2013). Or take, for instance, the media coverage of the mass protests in Delhi over the gang rape of a student, revealing how the public exercised its democratic right to assemble and express their outrage: "when it comes to grievances, India is a buffet. And anybody with a cause can find slogan-shouting time and space at Jantar Mantar—as powerful an advertisement for free speech as Speakers Corner in London's Hyde Park, only more crowded and more littered" (Lakshmi, 2013). Therefore, prodding into the past of urban parks in this book is not just about reminding us of how deeply entrenched the public political sphere is within our urban geography. Such an investigation also demonstrates how parks can serve as a meaningful and traveling metaphor in extending our political imagination across national terrains.

Furthermore, it is worth considering to what extent the nature of public participation in these material spaces has semblance to that of digital networks. In contemporary discourse it is claimed that while group-based 'identity politics' of the past were along conventional lines such as class, ethnicity, race, and gender, Web 2.0 architectures have fostered more diverse mobilizations. Nowadays it is claimed that individuals gather and activate around lifestyle values and engage with multiple causes:

Personalized politics has long existed, of course, in the form of populist uprisings or emotional bonds with charismatic leaders. The interesting difference in today's participation landscape is that widespread social fragmentation has produced individuation as the modal social condition in postindustrial democracies, particularly among younger generations. (Bennett, 2012, p. 22)

While indeed the technical affordances of social networks facilitate this process and allow for virtual corners on a range of topics from the profound to the inane, can this be a unique attribution to the digital sphere? As we have extrapolated earlier, because of their struggle for democratic architectures, urban parks have historically allowed for temporal social collectivities around issues that permeated conventional group identities, affiliations, and politics. To conclude, be it the People's Park, Beijing Park, or Hyde Park, there is a normative ideal or best imagined use of public leisure space endorsed by the state, imperial entity, or royalty that stands against the wide spectrum of social practice within these parks in urban societies. Paternalistic intent of the state or a private-sector interest often drives the design and shape of these public spaces, with a hope to convert the masses into modern, cultured, and active consumers of society. These ideals stand for aspirations and expectations, a powerful motif that gets transcribed and reified over time.

However, through ongoing interaction and participation of the masses, historically these public leisure spaces have morphed into emblems of freedom and human dignity. The continuous public struggle to democratize these leisure realms accumulates to form a rich social memory of these spaces, impacting future ideology and public protest. Of course, not all parks serve along the same lines of social activism, much like not all digital networks propel political participation.

That said, when we look at the history of these spaces and compare them across cultures, there is a critical relationship between public leisure sites such as urban parks and social protest that cannot be ignored. They serve as public platforms wherein a range of ideologies can play out. Within this social theater, democratic practices continuously emerge, in spite of architectural manipulations and surveillance infrastructures from above. This is much along the lines of Lefebvre's (1991) distinction between representational space (appropriations and usage of park space by the masses) and representations of space (design and control of park space by the authority). Such leisure platforms take on a hybrid identity where corporate branding, political campaigning, and propaganda battle it out. Yet, if we are to take a cue from the history of urban parks, the critical mass that harnesses these leisure spaces for political activism demonstrates a human persistence with a capacity to transform a typical urban park into a protest space.

Creative and Playful Protest

"It should be clarified that a new public space is not synonymous with a new public sphere," remarks Papacharissi (2002, p. 11), reminding us that it is the nature of social mediation and interaction that makes a space 'public' and not just its underlying architecture. Just because social technologies create the fabric of democracy, it is still contingent on user activities to materialize such ideals. If you build it, it does not mean they will come. And when they do, it will continue to be a challenge to keep them engaged and dedicated to the cause. So, not surprisingly, activism does require abundant creativity to keep the momentum going, particularly among a disparate populace. To convert passion into sustained action, it often helps to have a sense of community. In recent years scholars have found a range of ways in which communities of protest are being brought together, one such trend being that of 'fan activism.' This is a phenomenon where protesters appropriate fandom practices for their political movement:

> Scratch an activist and you're apt to find a fan. It's no mystery why: fandom provides a space to explore fabricated worlds that operate according to different norms, laws, and structures than those we experience in our 'real' lives. Fandom also necessitates relationships with others: fellow fans with whom to share interests, develop networks and institutions, and create a common culture. This ability to imagine alternatives and build community, not coincidentally, is a basic prerequisite for political activism. (Duncombe, 2012, p. 1)

Specifically, popular culture can serve as powerful communal glue for a wide-ranging audience. Whether we are in Cape Town or Texas, people are well versed in regard to certain media icons and characters. Often, these figures are harnessed to weave rhetoric of commonality and solidarity. In her classic text, *Entertaining the Citizen: When Politics and Popular Culture Converge* (2005), Liesbet van Zoonen brings to our attention the playful, affective, and fantasy aspects of fandom and how this impacts political discourse. She reminds us of the inherent imaginative properties of popular culture that stimulate intensity and inspiration within the masses, creating an emotive bond. This bond has exponentially spread with the deployment of social media. This underlying affective thread and common reference point can be useful for activists promoting social change by deliberately tying popular media rhetoric to their political cause. For instance, much loved *Star Wars* images, such as lightsabers, stormtroopers, and Princess Leia, kept surfacing during the Occupy Wall Street movement in Zuccotti Park.

Here we see how mass protest and popular culture are natural allies in the political sphere. Before we start to romanticize this notion of empowerment through popular culture, there is evidence on how this can also serve the interests of corporations and the state. It is often just a matter of time before popular concepts such as flashmobs become fodder to sell more phones—think of the T-Mobile YouTube viral campaign. In the words of Mark Dery, a cultural critic and blogger, "the look and feel of culture jamming, at least, have been appropriated by the mainstream, tirelessly promoted . . . and hijacked by guerrilla advertisers to ambush unsuspecting consumers" (2010). We see a similar pattern in political campaigning from the United States to Germany where flashmobs, online posters, and humorous slogan competitions via online social networks have shown successful results in reaching their target audiences.

If we trace our path back to early 19th-century Beijing or to the London parks, protesters were just as innovative in their usage of space, forming human chains, holding humorous signage, at times dressing in costumes, and using theater to sustain their audience (Gough, 2000; McKay, 2011). These activities aimed to create temporary solidarities, transforming an abstract mass into a united civic group that shared common political concerns. Mass public entertainment was a way of communicating efficiently across a diverse public, unifying and making visible common messages directed at the authority of concern. In the late 19th century, in cities across the United States, a variety of groups used public parks to stage parades, heritage celebrations, and rallies with placards poking fun at the authorities as a means of expression that superseded ethnic, racial, religious, and sexual identities. Given the inherent challenge of creating a community feeling among a disparate group across class, gender, ethnicity, and religion, playful and creative means of communication became essential in the formation of mass protest. What was seen as effective was to capture "secular ritual forms which express communal values and sentiments by symbolically abstracting features of the social and normative structures from which they derive" (Lawrence, 1982, p. 155).

When discussing creative public subversion in a historical context, it is impossible to not evoke the eloquent work of the Russian literary master Mikhail Bakhtin. He delved into the popular carnival practice during the medieval period and its centrality to social life during that time. He encapsulated this event as the 'carnivalesque,' the temporal licensed anarchy in the town square that served to foster communal feeling of unity and belonging in an era of deep social division. He argues that in the long run, the disruption of social order is ironically essential to its maintenance:

> The suspension of all hierarchical precedence during carnival time was of particular significance. Rank was especially evident during official feasts; everyone was expected to appear in the full regalia of his calling . . . and to take the place corresponding to his position. It was a consecration of inequality. On the contrary, all were considered equal during carnival. Here, in the town square, a special form of free and familiar contact reigned among people who were usually divided by

the barriers of caste, property, profession, and age. The hierarchical background and the extreme corporative and caste divisions of the medieval social order were exceptionally strong. Therefore such free, familiar contacts were deeply felt and formed an essential element of the carnival spirit. People were, so to speak, reborn for new, purely human relations. These truly human relations were not only a fruit of imagination or abstract thought; they were experienced. The utopian ideal and the realistic merged in this carnival experience, unique of its kind. (1984, p. 10)

Bakhtin notes that this political performance has persisted and preserved in modern forms of the theatrical and the spectacular to ignite our innate need to belong to a larger whole. These humorous ploys, when reenacted time and again, can become a cross-cultural brand in itself. For instance, after the 1986 Chernobyl nuclear disaster, a group of older women exasperated by the current apathy of the government, decided to put on a farcical public protest by calling themselves the 'Raging Grannies.' Capitalizing on the stereotype of older women as sweet, silent, and to a large extent invisible, these ladies occupied the public square adorned in playful costumes that deliberately challenged the cliché. They came with outrageous hats, gaudy shawls, frilly aprons, and pink running shoes; some wore long white gloves and patent leather purses to feed the expectation. This overall spectacle commanded the attention of the passersby without being threatening. And with this captured audience, the Grannies usurped public leisure space for their political agenda and, to this date, continue to inspire protest in their own unique way.

Today, more than seventy dynamic groups of Raging Grannies are busy raising a little hell for authorities across Canada and the United States, and in Japan, Greece, and Israel. While there is no central organization and each group has autonomy to focus on its own concerns, there is a network established through Internet and un-conventions held every second year. This, then, is a network made resilient by local autonomy yet strengthened by the exchange of views, ideas, and songs through the Internet . . . This network of powerful women not only is challenging authorities and stereotypes of older women, but also broadening the notion of what it means to age while reinventing protest. (Roy, 2007, p. 151)

We see the Raging Grannies protesting at Liberty Square in New York and the Freedom Plaza in Washington, D.C. (the plaza renamed in honor of Martin Luther King Jr. and based on his 'I Have a Dream' speech). Postings online go up frequently to recruit participants for the Raging Grannies. For instance, on February 23, 2006, a post went up on the social network site Yelp that announced the plan for a new Raging Grannies spectacle at

Justin Herman Plaza in San Francisco to protest against wiretapping. The description read:

> Plan to attend this fantastic event: Vigil: Stop Wiretapping Now!, a Constitution vigil in solidarity with a nationwide http://MoveOn.org action.
>
> If you can, Please make signs that say "Bush Broke the Law" and "Stop Illegal Wiretapping" (it's perfectly okay to come without a sign too)
>
> The visual is great: an ailing Statue of Liberty (we have a super costume ready) will be played by one of the Raging Grannies. The rest of us who want roles will [be] the medical team with the cure presenting our prescription to Congress Rx: Independent investigation of wiretapping to restore justice. If you want to look authentic for the camera (this event will be videotaped and posted in Indybay) come in a medical personnel costume . . . OR just come as yourself and join the Raging Granny "mourners." To make a faux prescription on a postcard size paper, just write "Prescription" on the top and hand write "Rx: Independent investigation of wiretapping" and you can sign your name and address at the bottom. We'll bring extra materials so you can make a card at the event. THEN some of us will deliver all the prescriptions to Rep. Nancy Pelosi on Thursday, Feb 23.
>
> Bring a flashlight for the vigil.
>
> Singing will be lead [*sic*] by members of the San Francisco Raging Grannies and the renowned Peninsula Raging Grannies Action League. Please tell people about this event and ask them to sign up at http://political.moveon
>
> Let me know if you have questions and/or ideas for how to make this a powerful, media covered event. Visit our website: http://www.RagingGrannie . . . Thanks, Rev. Judy info@RagingGrannies.com

While this undoubtedly draws attention to the cause and visibly demonstrates against the government, it also benefits the city council's agenda for increasing tourism and sustaining property value in this much mythicized city. After all, it falls well within the social imagination of what constitutes San Francisco and legitimizes the city's historic position as antiestablishment. Arguably, this is needed more than ever with the current 'yuppification' of the city, with dot-com billionaires and Silicon Valley techies yearning to believe that the Californian Ideology is still very much alive and thriving.

Not all governments view these 'tactical frivolities' as useful to social functioning. There are times when protesters are faced with a humorless state and have to change their strategies to be at once visible and invisible, humorous and yet straightforward to the authoritarian eye. Historically, the more authoritarian a regime, the more critical humor has been in carving out public protest:

These protests, history suggests, are particularly prevalent when those benefiting from rampant political corruption lose their sense of humor, become ridiculous in their seriousness, but are incapable, for one reason or another, of silencing their prankster publics. There would appear to be important and ongoing tensions, then, between the shifting humors of state agents and the productive capacities of critical citizens, suggesting that a fuller appreciation for the dynamics of those tensions is an important step in understanding how challenges to power can result in positive political change. (Bruner, 2005, p. 136)

Thereby, the type of state affects the nature of political performance: 'Prankster publics' turn to a wide range of public expressiveness from the carnivalesque to the subtle. Take the case of China and the creative modes of protest practices within public leisure domains. As mentioned earlier on, urban parks in China are open yet deeply regulated, particularly when it comes to mass gatherings. However, there are gray zones sanctioned by the state that allow for crowds to amass for social purposes, such as the popular *qigong*, healing through breathing exercises (Davis et al., 1995). For this activity, groups engaged in this pursuit are allowed to print and distribute flyers and affix signboards and public announcements. Historically, *qigong* is a practice that has deeper meaning. During the Mao era, it was considered a way to foster private mental spaces in a public setting and herein, "urban constructs of parks give way to private experience" (p. 359). *Qigong* has a long tradition of association with peasant uprisings and heterodox movements, such as the Boxers, who practiced *qi* exercises and promoted their visions of a utopic society. The popular image of *qigong* founded on media and healing narratives has created a sense of autonomous identity that is well entrenched in urban spaces and the social imaginary. It has unsettled the Chinese state, as it is perceived to have political and subversive potential. Hence, there is a continuous struggle between the state and the public on what are deemed as public leisure enactments:

People take up *qigong* because of disenchantment with official ideology and policy. The state's presence is inserted into everyday life through surveillance of public arenas such as the parks. Categories of 'official' versus 'false' *qigong* are created to permit practitioners of 'superstitious' activities to be taken into custody for questioning. Those who continue to practice in parks do so under red banners and white certificates of legitimately recognized schools of *qigong*. Witch hunts of masters are carried out in the name of corruption. And boundaries of normality are reestablished through creating a medical disorder called *qigong* deviation. (p. 360)

This pattern of weaving discontent into the larger matrix of leisure is evident in the virtual sphere in China. The top 10 Internet activities in China are listening to or downloading music, reading news, using a search engine, instant messaging, playing games, watching videos, using a blog or personal space, emailing, using a social networking service, and reading Internet literature (Wallis, 2011). Chinese cyberspace is mainly perceived as a place for socializing and entertainment as users describe their web-based activities as 'fun.' Yet research has revealed that within this innocuous maze of public leisure, we can see the emergence of diverse and creative communicative codes and homonyms that can only be understood by participants that share interest in a common cause. For instance, there is a gay realm in Chinese cyberspace that is socio-linguistically constructed through terms such as *tongzhi* or 'comrade' (a euphemism for gay or lesbian) and other inside literature. To Giese, "the real subversive potential of the Internet in China arises not because BBSs (and blogs) are used for overtly political expression, but because of the anonymity, freedom of expression, and opportunity for negotiating identity that such spaces allow" (2004, p. 23). For example, in 2003 the Mu Zimei phenomenon began, where a young woman in Guangzhou stirred up controversy when she began blogging about her active sex life. Such blogging rejected conventional notions of romantic love and served as a channel that opened up conversation on issues beyond sexual politics. Yang (2009) has argued that the "personal is political" in China, although not in the manner in which 'Chinese politics' is usually conceptualized in the West.

Furthermore, Chinese users who want to express views the government might frown upon have developed technological solutions like VPNs and anonymizing tools, employing software that changes the direction of text as well as non-technological methods to get their messages out:

> For example, after the July 2009 riots involving Muslim Uyghur's and Han Chinese in Urumqi, the capital of Xinjiang province, all online discussion, photos, and video of this event were blocked. To get around such censorship, clever Internet users employed a practice called "tomb digging," or "digging up" earlier posts about Xinjiang or Urumqi, and then adding comments about the riots . . . Other means of avoiding censorship include using encoded language through relying on the use of oblique references and metaphors, and through taking advantage of the richness of the Chinese language, with its multiple homophones. Still another practice is to split headers in an otherwise blank posting so that the user has to pull the pieces together . . . Through these and other creative techniques, Chinese cyberspace has become a realm for polyphonic expressions to exist outside the dominant discourse, and as such, it is constitutive of social change in China. (Wallis, 2011, p. 422)

In fact, recent new media architectures have given birth to the practice of *e'gao*, a combination of the words 'evil' and 'to make fun of' that signifies a multimedia expression that spoofs or pokes fun at an original work (Jiao, 2007). Through practices such as Photoshopping images, creating lip-synching videos, or parodies of famous films, *e'gao* has succeeded in appearing with little agenda and yet has scripted a form of public resistance. Such forms of implicit protest through creative mash-ups have posed an ongoing challenge for the Chinese state as they are woven deeply into corporate platforms that benefit from such group participation and enactments. Hence, to a certain extent, the deeply commercial aspects of these infrastructures protect the public from state censorship.

To conclude, boundaries of inclusion and exclusion surface through the architecting, regulating, and mediating of public leisure space by those in authority, making visible the rights and status of individuals and groups. Yet, in practice, communities create novel modes of communicative practice that carve out spaces of political expression. Here, the state and the corporate sector hold the complex and often contradictory position of being complicit in regulating their citizens and yet, creating web architectures that commercially feed on such mass innovation and creativity that is, at times, implicitly political.

Overall, we see much research done on social network sites, particularly their potential to facilitate democracy through mass protest. Substantive work has focused on their evolving techno-social infrastructures and practices, compelling researchers to emphasize the unique spatiality of these virtual leisure platforms. When we make historical comparisons, it is usually between old and new media. Hence, to lend a fresh perspective to this popular field of new media research, this chapter has provided insight from a seemingly disconnected academic discipline—park studies.

Given the shared rhetoric between urban parks and social network sites—of being open, free, and democratic—this chapter has initiated a dialogue between these two fields to lend a more comprehensive perspective on the relationship between public leisure space and political communication. By looking at case studies of urban parks and social media platforms in the United States, the United Kingdom, and China, this chapter has argued that public leisure domains are ideologically driven and symbolically marked often by state and/or corporate agendas that can be seen as authoritarian, paternalistic, or libertarian. Oftentimes, we see openness as deliberately architected into these spaces to serve as a safety valve and a concerted effort to gain state legitimacy through a modern public image. While these spaces have been designed as means of containment of mass politics, they often can serve as a hotbed for protest.

Over time, such 'safety valves' can become powerful pipelines for social movements that span across groups and social contexts through grassroots agency. By examining the range of political communication within these public domains, we see that they are interwoven strategically and deeply within acts of leisure, often concealing them from state surveillance. Also, such playful, creative, and humorous modes allow for the permeation of social causes across conventional group affiliations and help form temporary social collectives that share a common cause. Basically, commercialism serves as a double-edged sword: it exploits protest for private gain and at the same time protects activism from being easily controlled. Overall, public leisure landscapes within the digital and material sphere share common agendas and architectures that, when viewed as a comprehensive and historically embedded space, give insight into the nature of political participation and mass protest.

4 Walled Gardens
Online Privacy, Leisure Architectures, and Public Values

Before I built a wall I'd ask to know
What I was walling in or walling out.

Robert Frost, *Mending Wall*

Garden images have articulated powerful notions of cultural and social identities, of who and what is to be included or excluded . . . The garden itself, part of a network of walled gardens, is traditional and permanent with its high, solid walls. The flowers found and sown within its walls are representative residents of domestic and social stability.

Mandy S. Morris, "'Tha'lt be like a blush-rose when tha' grows up, my little lass': English Cultural and Gendered Identity in *The Secret Garden*"

There is no overarching conception of privacy—it must be mapped like terrain, by painstakingly studying the landscape.

Daniel Solove, *Understanding Privacy*

Walls can protect as well as suffocate. Containing a space can be akin to sustaining a particular cultural mode of being. This is especially true as these enclosures become normalized over time. In such a context, the garden can be viewed as a social response to containing the wilderness, either by taming the undesirable or walling off from it. If we pry into the history of garden making, we discover the breeding of patriotism, class, and femininity manifested in a spectrum of controlled landscapes. Take, for instance, the early 18th-century English landscape and the obsession with the garden. This greening of urban space was a reverberation of the political climate that was consciously moving away from the then French 'dictatorial system' to a more 'modern' and flexible 'English' way of designing leisure grounds (Hunt & Willis, 2000). The walling of these greens was perceived as a more benign and enlightened sort; not the heavy-handedness that came from the princely states but rather, from a more nuanced and open perspective. According to Stephen Switzer, a key spokesperson for the early English gardens, it was desired that "where possible, enclosing walls should vanish, and by means of 'an easy unaffected manner of Fencing,' it would look as if the adjacent Country were all a Garden" (Hunt & Willis, 2000, p. 2). This was espousing not the removal of barriers altogether but the removal of the feeling of boundaries that came with a historical connotation of imperialism and monarchy. Humphry Repton, one of the last great English landscape designers of the 18th century, was instrumental in communicating these ideas through the manufactured greens:

> Repton's own fundamental contribution to landscape history was to reclaim gardens for social use and relate them again to the houses they served. He pushed back the park and reintroduced regular and architectural forms—terraces, raised flowerbeds, trelliswork, conservatories. These were the logical extension of the social spaces of the house. (Hunt & Willis, 2000, p. 32)

So while private gardens were embracing the aesthetics and values of openness and accessibility, there was a parallel trend of the closing in of the walled gardens during the Victorian era that intended to protect, preserve, and nurture. The initial design of these structures was driven by the main goal of producing more agreeable conditions for the select landscape to prosper and thrive (Campbell, 1987). Soon enough, however, this 'walled garden' perspective extended to the carving of social space along the lines of class, as well as marked spaces of protection for women and children. This came with a paternalistic rationale of security and safety, of preserving innocence and virtue. This meant there was a concerted recognition of what was undesirable and thereby, gardens were walled off from select territories. Over time, this concept of the 'walled garden' has taken on several guises, from the most confining to liberating to radically transformative in nature. In this chapter, we deal with four urban leisure landscapes: gated communities, shopping malls, children's playgrounds, and guerrilla gardens. Each form comes with its own (and often paradoxical and opposing) rationale, design, cultural practice, and gatekeepers, but all share in the struggle between the private and the public sphere and between inclusivity and exclusivity.

The focus is on walled gardens because the term has gained significant traction in the digital sphere in response to contemporary divides transpiring online. A 'walled garden' in tech speak refers to a closed or exclusive set of information services provided for users. At times, this message is sugarcoated as a necessary means to protect users from outside negative influences like online viruses and spam. Joseph Turow, the author of *Audience Construction and Culture Production: Marketing Surveillance in the Digital Age* (2005), points out that many online practitioners have begun

Figure 4.1 Security spikes on the fence of a gated community. *Source*: Edward/ Wikimedia Commons/Public Domain.

engaging in a 'walled garden' approach to the digital world, that being "an online environment where consumers go for information, communications, and commerce services and that discourages them from leaving for the larger digital world" (p.116). This is in contrast to giving consumers unrestricted access to applications and content as in the early era of the Internet. In other words, a 'walled garden' is:

> a type of IP content service offered without access to the wider Internet: for instance, most mobile telephone networks provided walled gardens to their subscribers. This has wider regulatory implications, involving the development of 'gatekeepers' rather than open access models. (Jacob & Walsh, 2004, p. 26)

Today, walled gardens have become the norm in our digital communications as we acquiesce to the e-reader permit within the Amazon ecosystem, ask for approval for downloading an application on Apple iOS, and are compelled to agree on multiple restrictions within diverse social media platforms meant for leisure. This trend is impacting those well beyond the economically resourceful. In recent news, David Talbot for the *MIT Technology Review* remarked that the digital divide has taken on a new walled garden form, led by technology moguls such as Facebook and Google. Platforms like the new Facebook Zero and Google's Free Zone are able to include a larger and less prosperous newbie clientele into the virtual leisure domain. However, the hierarchy of access to these enclosed structures continues:

> Victor Ferraro Esparza, a strategic product manager at Ericsson, maker of the network equipment that keeps customers of Telkomsel and other carriers within their walled gardens, says that the plan has led to an increase in wireless data subscriptions. 'What we saw there is that when they were simply offering an undifferentiated mobile broadband subscription for a monthly payment, they had relatively little interest,' he says. But once the carrier offered lower-priced special purpose plans— the first of which was Facebook Zero, plus a few chat sites—'they had a big increase in new subscribers, and were later able to convert some of those subscribers to a higher-priced unlimited mobile broadband package.' (Talbot, 2013)

Besides these commercial entrapments, there is the walling in and fragmenting of groups online based on their political, social, and cultural interests and commonalities. There is a marked shift from open and public sites such as MySpace to quasi-public digital social grounds such as Facebook (Boyd & Ellison, 2007). Niche social network sites such as BeautifulPeople and BlackPlanet have emerged as exclusive hubs of leisure, defined by appearance and race, respectively. Take, for instance, ASmallWorld, part of a breed of social network sites that sprung up in mid-2000 for just the wealthy. Today, it has grown to include 800,000 members globally with 65 percent of the members in the U.K., Italy, Germany, and France and 20 percent in the U.S. Currently they are no longer accepting members so as to maintain their exclusivity. According to Nicola Ruiz (2008), writing for *Forbes*, class does matter within the Web 2.0 leisure terrain:

> Though it once seemed that the Web was the last place where status didn't matter, the elite are now looking for a comfortable place to mingle with like-minded people. They're leaving Facebook and LinkedIn to the riffraff.

Sometimes walling-off takes on a subtler disguise although it can have just as powerful an impact. Within Facebook, the mere clicking on the popular 'like' feature fosters a segregation of taste and lifestyle. This often innocent act filters our affiliations to create a homogenous leisure platform that intends to foster "a common understanding of the world, a shared identity, a claim to inclusiveness, and consensus regarding the collective interest" (Livingstone, 2005, p. 9).

That said, we witness continued resistance and efforts to revitalize the Internet with alternative models of digital leisure space. Some examples include open source platforms, non-corporatized digital commons, the rebirthing of anonymity and intractability, and efforts of joint governance and communal sustenance. Vigorous debates pervade over the governance and architecture of Web 2.0 leisure geographies, particularly in the areas of privacy, property, paternity, and profit. At the heart of these discussions is the growing concern over whether we are losing the battle in maintaining the digital public sphere as a non-commodified and unified domain and if indeed this is the fate of the urban leisure commons.

In this chapter, we start by examining this ideal state and the underlying assumptions driving this vision. We draw on parallels between concerns regarding digital leisure space with that of material space, such as gated communities, shopping malls, children's playgrounds, and guerrilla gardens. Specifically, we look at the democratic implications that result from the gating of public space. We examine the architectures of commerce, security, and fear and their effect on movement within such leisure grounds, and we explore the pursuit of idealism, public values, and human ingenuities in the transformation of fortified enclaves into an inclusive and communal space.

Basically, the metaphorical application of the 'walled garden' overlaps with the physical counterpart where a user is less able to escape this area unless it is through designated entry/exit points or if the barriers are altogether removed. Whether it is a physical or virtual walled garden, we can learn much about its ideology by examining its architecture and the range of practices within such spaces, sometimes because of these constraints and, at times, in spite of them. Through this comparison, aspects of inclusion and exclusion, commercialism, protection, activism, and the regeneration of public leisure spaces are illustrated. In using the 'walled gardens' metaphor, we need to ask who and what is being protected by these 'walls,' what purposes do they serve, and which actors participate in constructing and sustaining such gated structures. In what ways are these gardens 'protective enclaves,' 'safe enclosures,' 'pleasure gardens,' and 'fortified terrains'? Here, privacy is seen through the lens of accessibility, choice, and ownership and reified through certain architectural trends of public leisure space—gardens within gated communities, urban malls, playgrounds, and community gardens. The intent is to allow for a cross-engagement of discourse between these fields to shed light on tensions between the public and the private sphere. To put it more dramatically, this helps to unravel the "soul of cyberspace" (Kramer & Cook, 2004, p. 178) as trust and vulnerability become central in gauging the social conscience of public leisure space.

DIGITAL ARCHITECTURES OF PRIVACY

The architecture of Web 2.0 has awakened the classic notion of the public sphere. Web 2.0 has been popularly viewed as a public domain for all people to participate and contribute to the shaping of its digital space. At the heart of this social engineering is the intrinsic need for community. There is a need for people to connect, share, and network for common pursuits—regardless of geographic constraints—within what Howard Rheingold enthusiastically proclaimed as 'virtual communities.' Indeed at the onset, this information architecture was framed as the new digital commons, the new republic, and the place to give democracy another chance. As James Orbesen of *Salon* reminisces:

> The early Internet was certainly a different place. It seemed a time of unlimited potential, when the old barriers to communication and information were said to melt away like so much butter in the microwave. People would be linked in ways never seen before, all in a purely public and noncommercial space. Early analysts claimed that the old media conglomerates were going to be swept aside by a coming Digital Age. For those looking to the future, the Internet would be the democratic space since its underlying principle, the networked sharing of data, was inherently leveling, free, and transparent. (2013, p. 1)

Eli Pariser, the Internet activist and author of *The Filter Bubble: What the Internet is hiding from you* (2011), points out that far from the democratic ideal of a 'community' that new media platforms promised, we live in an age where our networks are carefully curated by algorithms that are programmed to expose us to mainly like-minded people. In other words, we live in a 'filter bubble,' an invisible information architecture. As people click on certain links, view specific friends on Facebook, choose movies, and select which articles to read and forward, they are bound together based on their online choices. While new media platforms promise an open and free space, in reality people are being closed off from diverse points of view, tastes, and perspectives. Basically, Pariser argues that "a world constructed from the familiar is a world in which there's nothing to learn . . . (since there is) invisible auto-propaganda, indoctrinating us with our own ideas" (2011; p. 22). After all, if we are able to segment the public based on stable categories such as class, gender, and political affiliation, it makes it easier for the state and the corporate sector to regulate and control this populace. Hence, there is a growing concern that Web 2.0 has escalated the polarization and prejudice between social groups by enabling online communities to remain within their own private spheres of belief. These enclosed spaces can be viewed as 'echo chambers' where certain perspectives can be reinforced through constant repetition, making it appear as truth (Jamieson, Hall & Cappella, 2008). This poses a serious problem when it comes to creating a genuine public sphere. The creation of a public sphere requires an empathetic understanding of society that can only come from exposure to those outside our comfort zone. Of course, this perspective undermines the agency of the individual and the fact that their choices also shape these enclosed arenas. These spaces, while dominated by corporate and state interests, are also grounds of engagement, connectivity, and creativity that continuously alter and challenge the existing parameters within which these social narratives play out.

Of course, these architectures do not spring up in a laissez-faire fashion but are driven by certain values, ideologies, and group interests that continue to battle. The focus on values embedded within technical systems is not novel but comes from a larger discipline of science, technology, and society. There are different dimensions to consider when examining values within web architectures (Van Den Hoven & Weckert, 2008). To start, there is the *technical mode*, where designers strive to balance efficiency and functionality with the larger legal and political system of the state. There is the *philosophical mode*, where the origin of the technology creator's values is brought to question and the range of consequences is made explicit both at a technical and material level. Here, less emphasis is on the design of the space and more on recognizing the gatekeepers and their influence on this space. This could lead to rationalizing these decisions or critiquing for further change. Lastly, at the *empirical mode* such experimental inquiry is essential to ascertain whether the design of a particular technical space embodies intended values.

When we talk of walled gardens, what comes to mind is the value of privacy. This notion is currently being fought on an ideological battlefield and its ramifications manifest in the shaping of architectures, both virtual and material. Few would argue today that corporate interests are barricading social media platforms and creating an archipelago of walled gardens. Robert McChesney argues in his latest book, *Digital Disconnect: How Capitalism Is Turning the Internet against Democracy* (2013), that it was just a matter of time before capitalism took over the Internet and converted it into a deeply commoditized and privatized space. He argues that while we see capitalism and democracy as synonymous, they are opposing ideologies that need to be divorced. In his view, capitalism subverts democratic space in favor of profit. While at it, nothing is sacred, including privacy.

> This situation results not necessarily from a conspiracy, but rather from the quite visible, unabashed logic of capitalism itself. Capitalism is a system based on people trying to make endless profits by any means necessary. You can never have too much. Endless greed—behavior that is derided as insanity in all noncapitalist societies—is the value system of those atop the economy. The ethos explicitly rejects any worries about social complications, or externalities. (p. 17)

The conversation continues as neoliberals point to this pattern as part of the free market ideal. On the other hand, libertarians place their faith in the creative ingenuity of collective human action and resistance, which better reflects the public demand for non-corporatized space. While undoubtedly the power of community can compellingly shift the direction of the

Internet towards a more open and democratic ideal, the forces of the state and corporate interests are always at play. One needs to question whether the wisdom of the crowds is always right, as the recent Reddit fiasco on the Boston bombings revealed. In the name of exercising the democratic right on open social platforms, innocent people were unfairly tagged as possible suspects, creating significant damage to their personal lives.

The fact remains that Web 2.0 is now a deeply valued property and several actors have a vested interest in the ways in which the 'wilderness of cyberspace' should be architected, zoned, and managed. Divergent as these ideologies may be, one thing for certain is that their architecture requires a connection to the physical realm. They are situated within protocols that operate in material space and possess spatiality that so far has been little explored:

> Architectures don't come in natural kinds. My point instead is the choice—that there is a decision to be made about the architecture that cyberspace will become, and the question is how that decision will be made. Or better, where will that decision be made. For this change has a very predictable progress. It is the same progress that explains the move to zoning in cities. It is the result of a collection of choices made at an individual level, but no collective choice made at a collective level. It is the product of a market. But individual choice might aggregate in a way that individuals collectively do not want. Individual choices are made within a particular architecture; but they may yield an architecture different from what the collective might want. (Lessig, 1999, p. 1411)

The dynamism of social media's technical affordances has accelerated the urgency for such a dialogue. With each affordance, there is a real threat that social practice can be changed radically and irrevocably, for example, in how privacy is exercised, granted, and gained. From a dystopic point of view, these new features could further the surveillance of our daily lives, building on existing architectures of control and naturalizing these boundaries of containment. Such infrastructures could mean ultimate territoriality with the global spread of imperialism across domains. However, if we are to take the opposite perspective and embrace a more utopic lens, we are reminded of the fact that this narrative is not yet over. Historically, the advent of the World Wide Web has offered new opportunities for interfacing and community building. In this sense, it was time and again able to overcome the age-old stranglehold of corporate monopolies and the nation-state. This is reflected in one of the most important designs of web architecture, the URL:

> One goal of the Web, since its inception, has been to build a global community in which any party can share information with any other

party. To achieve this goal, the Web makes use of a single global identification system: the URI. URIs are a cornerstone of Web architecture, providing identification that is common across the Web. (Jacob
& Walsh, 2004)

However, when life plays out, we are neither living in a utopic or a dystopic
world but rather we face the deep complexities of the present social context.
Helen Nissenbaum argues in her book, *Privacy in Context: Technology,
Policy, and the Integrity of Social Life* (2009), that privacy preferences are
dependent upon context: "people don't choose in the abstract, but in a
particular context" (p. 105). She goes on to make the point that if we are
to judge the issue of privacy on social networks, we should examine the
context to determine public or private allowances. Again, the spectrum of
an event is laid out before us, shifting the meaning of what constitutes as
private or public: "we do not have a dichotomy of two realms but a pano-
ply of realms; something considered public in relation to one realm may be
private in relation to another" (p. 215).

Does this mean that we cannot talk in broader terms about public and
private architectures? When behavior is contextually influenced but recurs
time and again, does it not become normalized and built into the culture of
that specific space? Does this approach give unrealistic power to individual
choices and less to larger sociopolitical systems? Alan Westin (2003), a
professor of public law and government, talks in broader terms in regard
to privacy and freedom, remarking that times have changed: individuals
are now more willing to divulge their personal lives. More so, the social
fabric has also changed to become a more permissive society. According to
Westin, one thing seems to have remained the same: the element of curios-
ity that drives online communities to pry and prod through social network
sites for private information. He provides a framework to examine how
privacy norms are set in society through the three lenses of the political, the
sociocultural, and the personal. For instance, privacy at the political level
must be considered within the larger system. For example, an authoritarian
society would see the public sphere as central to social life and would view
privacy as an antithetical requirement. On the other hand, a democratic
system would favor individual freedom and expression but, at the same
time, within this system the 'free market' would be given much weight as
a deliverer of social progress. At a sociocultural level, class and race factors do influence the degree of freedom to which one has access. What is
'personal,' how is it exercised in public space, and how is its position as a
public good dependent on the social norms of the time? Lastly, the claims of
privacy are also exerted at the individual level, where individuals differ on
the extent to which they want to disclose and communicate to the public.
Of course, this self-management is deeply influenced by the political and
the sociocultural but is also to some degree independent from the former
categories, resulting in a diversity of enactments.

We indeed have come far from the *Declaration of Independence for Cyberspace*, where Barlow told governments to "keep your hands off the Internet" (1996, p.3). Spaces, however, continue to be encrypted through rules of access and use of such 'public' property. We find ourselves subject to more rules and yet we volunteer to be regulated. Such an environment can be defined by its 'high plasticity,' which Greenleaf (1998) points out is an essential characteristic of a vibrant architecture seen in both physical and virtual spaces. Here, flexibility is itself a value and plasticity can be inscribed in the planning stages of our environments, promising a more responsive approach to changing value systems.

FORTIFIED ENCLAVES OF THE PAST AND PRESENT

In this section, we look at some key arenas of the walling of public leisure space that elicit privacy and examine the discourses and practices surrounding the virtual and physical architectures. There is much concern regarding these spaces because there is fear that these domains signify a retreat from civic and democratic life and a reinforcement of social fragmentation. While we witness deep cleavages along lines of gender, class, sexual orientation, and nationality attributable to biased architectures and practices, we also recognize continuous resistance, appropriation, and play with these spaces, both virtual and material, and both historical and contemporary. We will demonstrate that while the fragmenting of groups can be antithetical to the public leisure commons, it can also provide secure havens and intimate enclaves of sociality.

Democracy behind Walls

The ideal of an 'open' public can be perceived as naïve when society naturally segments into social enclaves. As society scales and expands, there is a need to reinforce community and that sentiment prevails in choices of more quasi-public architectures. Moral panic ensues when it is perceived that these communities are retreating from the public sphere and into defensible and discriminatory spaces, abdicating civic responsibility (Hunter, 2003). More importantly, these semi-private leisure enclaves are seen to threaten the sacred notion of democratic space such as the urban park, a landmark achievement of the public domain. The walled gardens appear to threaten the common good with the rise of private interests exercising their power within the public landscape. The urban park was a brilliant innovation and materialization of 19th-century idealism of an inclusive public space. Yet, in practice, these domains often excluded certain groups, such as the poor, the homosexuals, and the blacks, and stringently regulated the movement of women and children.

In the *New York Times*, Charlotte Devree (1957) calls Gramercy Park in New York City "a Victorian gentleman who has refused to die" and for good reason. Built in 1833, this private park on East 20th remains as one of the few parks in the United States that serves as a walled garden. According to John B. Pine's 1921 book, *The Story of Gramercy Park*, this park was segmented by class: To access it, you needed a special key which was given to only trustees who lived in that area. Key or no key, there were definite barriers to accessing these supposed public parks. The Victorian influence in fact pervaded across the United States, where the urban park, while intended for the public, was implicitly defining what constitutes as the acceptable commons; the poor, the immigrants, and the homosexuals were actively discouraged from participating within these public enclosures. This segmentation was hardly limited to the United States. We witness the underlying marginalization of homosexuals in public plazas in Moscow and Rio de Janeiro towards the end of the 18th century (Higgs, 1999). The boulevard ring of the Moscow plaza served as a main point for socializing during the post-revolutionary era, allowing little room for the poor or the homosexuals. In Rio de Janeiro, public gardens were designed and socially marked for strolling by the higher echelon of society, much like the southern European custom of family promenades that served as a way to inspect other families for potential marriage.

However, let us not undermine the agency of marginalized groups in exercising their needs and demands onto these socially-biased engineered spaces of leisure. The story is more complex than simply social control or mass upheaval. Be it the Moscow plaza or the Rio de Janeiro gardens, homosexuals leveraged these fabricated designs and movements to embed their own agendas. This was even more remarkable in a day and age where homosexuality was an illegal act and did not fit the acceptable performance within a public space. For instance, the family promenades provided a cover for gay glances and became one of the main outlets for gay cruising. This gave birth to a new way of expression that was erotic and semiotically coded, enabling a newly invisible class of marginal publics to perform within a heterosexually marked space. Examples abound of such communicative play while cruising across parks. For example, in the United States, the question, 'Are you a friend of Dorothy?' was a euphemism used to gauge sexual orientation without the general public knowing its meaning. Similarly, in England they would question if you were a 'friend of Mrs. King,' thus creating a signal of common acknowledgment.

Sometimes, these small acts of appropriation add to a larger act of territorializing. To illustrate, take the example of the Ramble in Central Park in the early 20th century (Rosenzweig & Blackmar, 1992). Frederick Law Olmsted and Calvert Vaux, the visionaries of this green space, believed that New York City needed a secluded landscape that was less manicured and urbanized and more 'natural.' This 'wild garden' was designed deliberately with a rough topography of woodlands, deep coves, dense planting, and rocky pathways, discouraging the regular park traffic of carriages. Interestingly, this very architecture formed an ideal setting for homosexual encounters and by the 1920s, to the great dismay of the mayor, this lawn came to be seen as the 'fruited plain.' These 'antisocial' acts succeeded in taking over a segment of Central Park for their own making in spite of the elaborate efforts to nurture a space for families. To this date, the Ramble continues to serve as an icon for outdoor gay sex, signaling the long-lasting impact that the persistence of subversive acts can have on such leisure domains.

Likewise, initial expectations of social network sites saw them as new democratic public spaces that fostered relations across conventional social boundaries. Contrary to these expectations, recent empirical evidence has found that these spaces in fact perpetuate and strengthen existing social capital. In other words, far from creating new social ties, these digital leisure domains maintain traditional relations. A study by DiPrete and colleagues (2011) reinforced this outlook: They analyzed segregation in social networks based on acquaintanceship and trust. While there was hope that this digital public sphere would foster the bridging of social capital, they discovered that segregation ran deep along lines of class, race, political ideology, and other socially divisive factors. What was further disturbing was that integration across social groups was found to be more challenging as many people still "isolate themselves from others who differ on race, political ideology, level of religiosity, and other salient aspects of social identity" (p. 1234). Given that social integration has historically been the core focus for the design and architecting of public space, there is to date little hard evidence on the extent to which "[people] have contact with people who differ from themselves on core status and values dimensions" (p. 1235). On a more optimistic note, there was evidence of a growing circle of trust that may serve as a future facilitator to address existing social cleavages.

In addition, within these shared networking platforms, groups emerge along lines of sexual orientation, relationship status, and age. For instance, exercising sexuality ranges from the miniscule shifts in how many times one changes his or her profile picture to the nature of postings that people put on their own and other people's walls (Alpizar et al., 2012). Recent studies indicate that a male tends to more often describe himself as being single and frequently engages in entertainment-related activities. It was also found that users who reported themselves as being homosexual or bisexual would more regularly change their profile pictures and personal profile information and less often post comments on others' profiles. These social cues add up to a larger communicative code that is able to carve out its own private space. Take another example from John Edward Campbell's book, *Getting It on Online: Cyberspace, Gay Male Sexuality, and Embodied Identity* (2004). He argues here that gay men have created several walled gardens online for self-expression such as the gaychub (a community celebrating male obesity), gaymuscle (a community constructed around images of the muscular male body), and gaymusclebears (a space representing the erotic convergence of the obese and muscular male bodies emerging out of the gay male 'bear' subculture). Communication on these platforms is coded to create spheres of inclusivity and exclusivity, with humor playing a strategic part in segregating the insiders from the outsiders.

Rather than preoccupy oneself with the superstructure versus agency dilemma, we need to recognize that there is a complex interplay between diverse actors regarding the private and public aspects of leisure geographies, whether virtual or material. Take, for example, the historian Roy Rosenzweig's exploration of the Massachusetts parks that were initially designed to socialize and civilize the public via preferred norms of society. Instead, they "were providing a setting for precisely the sort of behavior they were supposed to inhibit" (1979, p. 40). In fact, the introduction of the parks did not "remake the Worcester working class in the image desired by the State, the industrialists and reformers; neither did it precipitate a new class solidarity or consciousness" (p. 42). If anything, it gave a certain amount of autonomy to varied individuals and groups to shape these leisure grounds in accordance to the needs of their community. At the same time, they were bound by their socioeconomic background, gender, and status. Their social positions served as a constraint on their social practice within these parks based on the amount of time they had for leisure, who they could spend their time with in public, and for what purposes. Legal sanctioning was imposed upon certain groups to manage their movements within these spaces: They were restricted on the grounds of loitering, idleness, and public celebration. However, it was hard to sustain these regulations as immigrants and the working class collectively pressed on with their presence within these spaces. Over time, their presence became normalized in spite of the initial intent by the state. Challenging simple rules such as 'Do not walk on the grass,' served as a small way in which their presence was felt. At this point, the state found it beneficial to look the other way; this created a safety valve within society and contained unrest. Ironically, the authorities began to find the masses beneficial for the maintenance of urban parks, as these spaces had become a common asset and value among the populace. This made it easier for the state to keep these parks secure and clean and to maintain them as places for socially appropriate leisure activities for families across different demographics.

Similarly, when it comes to regulating online leisure spaces, the 'builders' and 'regulators' of online platforms have found it beneficial to attend to the 'users.' At times, they have acquiesced to user requests, because their manufactured leisure domains gain credence through usage by a diverse and active public. Yet this relationship is in constant and unresolved flux as control for such spaces are contested and/or shared on an ongoing basis. For instance, controlling online leisure spaces takes place through contracts, Terms of Service, and End User License Agreements (EULAs) for participation. If not adhered to, the platform owners have full right to block or kick out those who violate these terms.

> Game designers and platform owners control what goes on in the virtual world in two basic ways: through code and through contract. First, they control what can be done in the game space by writing (or rewriting) the software that sets the physics and the ontology of the game space, defines powers, and constitutes certain types of social relations. Through code they can change features of the virtual landscape, grant or deny powers to participants, and kick participants out. They can also write the code to allow them to watch surreptitiously what is going on in the game space. Because they can magically change the physics of the game space and see everything that is going on there, the platform owners are sometimes referred to as the "gods" or "wizards" of the game space. (Balkin, 2004, p. 2050)

On the flip side, though, designers' freedom and players' freedom are often synergistic as the value of an online leisure space rests on its usage. It is to the designers' benefit to keep their users happy and pay heed to their needs. Often, when designers makes a decision that a good number of users are unhappy about, they find themselves revoking that decision: "many of the most important controversies in game worlds revolve around the potential conflicts between assertions of the right to design and counter assertions of the right to play" (Balkin, 2004, p. 2051). To further complicate matters, users become the designers themselves through content generation, altering the spatial culture of the inhabited game. The digital space is 'owned' to some extent by users as they create new social norms and cultural practices. Together, the platform owners, designers and the users play out their roles in orchestrating and sustaining the gaming terrain.

> In virtual worlds, the relationship between platform owners and players is not simply one between producers and consumers. Rather, it is often a relationship of governors to citizens. Virtual worlds form communities that grow and develop in ways that the platform owners do not foresee and cannot fully control. Virtual worlds quickly become joint projects between platform owners and players. The correct model is thus not the protection of the player's interests solely as consumers, but a model of joint governance. (Balkin, 2004, p. 2082)

Interestingly, most states have created laws to sustain such leisure spaces as open public forums, by upholding the right to freedom of expression. For instance, while the state is more preoccupied with indecency and violence, platform owners may be more concerned about engaging and garnering loyalty from their users. And users themselves may be most concerned about building relationships, entertainment, information gathering, or just the passing of time. In fact, the best form of regulation comes from the users themselves through formal and informal enforcement of gaming norms and they shun or reprimand those who behave disruptively within the game (Donath, 2007). Thereby, in considering the dynamics of democracy behind walls in digital and material gardens, it is worth keeping in mind that different public leisure spaces evoke diverse social performance, some more open than the others.

Papacharissi (2009) examined three social networks—Facebook, LinkedIn, and ASmallWorld—to understand how their architectural features influenced the makings of community and identity. Facebook was unsurprisingly the largest and most widespread virtual geography, consisting of 47,000 college, high school, employee, and regional networks, handling 600 million searches and more than 30 billion page views a month. LinkedIn came with a more modest 500 page views per month, ranked 193rd in Internet traffic, while the recently dubbed 'A Facebook for the Few,' ASmallWorld, caters to an exclusive audience, ranked 9,571th in position. By comparing these platforms, Papacharissi revealed a balance of the private and the public, unique on each of these sites. Aspects of self-presentation, the display of taste to signal sociocultural identity, and the degree of tightness of the settings play a role in the nature of these walled social structures. Facebook emerged as the most publicly open structure with a high potential for manipulation by its populace to craft their private spaces. On the contrary, LinkedIn and ASmallWorld were more tightly-bounded and administered in their settings, fostering closer allegiances and perpetuating more conformity among members, creating a carefully manicured cultural norm. We see human behavior is subject to the range of socio-technical architectures. Human behavior can shape and is shaped by ongoing social communication. Papacharissi suggests that while there is no inherent promise or predisposition to these virtual geographies, there are indeed dominant cultural modes to diverse public leisure architectures that are worth examining:

> Future studies of the architecture of online spaces could examine personal interpretations of the options provided, and analyze how individuals incorporate, reject or adapt the architectural elements suggested by a provider. Content analyses could consider the extent to which individuals conform or deviate from available templates and the resulting impact on the interaction sustained by the online social network. Moreover, ethnographies of online social networks would be integral to understanding how members internalize overt and subtle spatial and behavioral suggestions and how they in turn adjust their behavior. Certainly, each social networking site serves a unique purpose, so network architecture is essential to meeting these unique objectives. (2009, p. 216)

To summarize, walled gardens are simultaneously public and private. The nature of the private–public balance depends on the social enactments within these spaces and architectural and cultural constraints and affordances, as well as larger ideologies of democracy dictating these geographies. Boundaries between the private and the public in the digital and material leisure domain are not just architecturally construed but are also socially embedded, with or without the walls. In a broader sense, walled gardens are not reified entities. Barriers that mark social behavior and space are a product of complex negotiations between diverse actors. Discrete spaces can indeed be deeply empowering, as with homosexual practices within the Moscow and Rio de Janeiro gardens or with the poor and immigrants exercising their right to commune within parks in Massachusetts. That said, it is worth acknowledging that these spaces are rarely neutral; they are deeply value-laden geographies oriented towards the more resourceful, be it socio-culturally, economically or politically. These platforms often serve as a hotbed for a range of human ingenuity as they craft new subversive and playful codes of communication.

Gardens in Modern Security Parks

Let us shift our attention to a modern walled garden phenomenon that, in the name of privacy, is viewed as a retreat from the urban commons ideal—that of gated communities. These are privatized public spaces with designated parameters or 'walls' that control membership and movement within. The rationale driving this architecture is that public social space is unregulated and uncontrolled. It is believed that the public domain creates security issues and undermines communal relations and quality governance. Ironically, these "enclaves can attract consumers searching for a sense of community, identity and security" (Grant & Mittelsteadt, 2004, p. 914). In recent years, these gated communities have become the norm as we follow their growing presence across nations. Gated domains carve out leisure spaces in the form of playgrounds, semi-private gardens, and entertainment and sport centers. While indeed this new social architecting is internationalizing, its social fractures are unique to the context at hand. Umpteen examples of segregation pervade, such as gated communities structured along the lines of religion in Israel (Rosen & Razin, 2009), race in South Africa (Hook & Vrdoljak, 2002), caste in India (Falzon, 2004), and urban inequality in Shanghai (Pow, 2007). Hence, there is a strong argument challenging the notion that gated communities represent a communitarian ideal or private choice that lacks wider social repercussions. Here, these architectures are accused of insulating these groups against perceived 'unwanted encounters.'

Basically, the formation of gated communities has given rise to strong critique where they are seen as symptomatic of the imminent breakdown of society, culminating in these fractured social enclaves and the retreat of citizens from meaningful public spheres (Low, 2003). By and large, there is common agreement among urban geographers that this growing spatial formation fosters social tension. Excessive encroachment by private forces on public spaces undermines traditional forms of urban integration and, ultimately, a meaningful public life. The harshest critics call this an 'urban pathology' and 'splintering urbanism' of this day and age (Graham & Marvin, 2001). Overall, actors that participate in the shaping of these architectures—designers, developers, residents, and the municipality, or basically the state, market, and civil society—are beseeched to consider questions of the potential trade-offs when privatizing public space:

> Is the project self-governing or a residential enclave within an established political unit? How does the settlement integrate with the larger community? What impact will gating have on the larger community? How might it affect traffic and crime patterns? Will it contribute to urban fragmentation, social segregation, and perceptions of crime? Because communities find different responses to these questions, their approaches to gating vary. (Grant & Mittelsteadt, 2004, p. 926)

Hook and Vrdoljak (2002) point our attention to the "spatial logic of apartheid" (p. 217) that permeates the design of gated communities in South Africa, where these 'security-park' complexes insulate and divide groups based on race. This ideology can filter down to ever more micro-levels of exclusion, separation, and division. Much of this discourse on security and safety feeds into the design of these "architectures of fear" (Ellin, 1997, p. 5). Living behind these walls is seen to promote fear of the unknown and foster a more inward-looking public.

Another example of division is drawn from the case of Israeli gated communities wherein barriers between Israeli and Palestinian communes are seen as more than just a security measure. In fact, a high court ruling suggests that these walls of development should not be reduced to security concerns but should take into account aspects of human rights (Rosen & Razin, 2009). Shanghai provides for another situation of segregation wherein 'civilized enclaves' are justified along the lines of urbanization and modernization (Pow, 2007). This "moral ordering of urban spaces" (p. 1539) is fundamental in shaping territoriality and exclusion between city dwellers and peasants in Shanghai's gated communities. Such discourses are masked and depoliticized through the lens of modernization and the 'civilizing' of the populace. The postmodern political geographer Edward Soja (2000) notes that such collective measures of gradual servitude to authority will enable private governance to play an increasingly public role. After all, public space ideally is meant to contribute to egalitarianism and liberalism by fostering unplanned social interactions between individuals who would not otherwise associate with one another. However, as public goods are encroached in the name of privacy, we can lose not only public space but also the "opportunities for shared experience and positive interaction among individuals from diverse backgrounds" (Caldeira, 2000, p. 306). Thereby, we can make the case that design principles of gated communities are antithetical to what constitutes an open and unmediated public domain. In other words:

> gated communities appear as segregated spaces with a social ecology that is planted into the fabric of the city; where the wall starts a new social area begins, whether one lives inside or out. (Atkinson & Blandy, 2005, p. 180)

This conversation on the gating of public architectures and its negative consequences on social values finds its echoes in the virtual terrain. As explained earlier, Eli Pariser's 'filter bubble' concept of the gating of social network space has raised the alarm on the direction in which our society is heading, shaped by these powerful means of digital segregation. These algorithmic walls are programmed to expose people to others who are mirror images of themselves, sharing similar interests, political affiliations, and socioeconomic status. Therefore, these new info-structures appear to threaten the democratic ideal. This perspective has been vociferously pushed by the legal scholar Cass Sunstein, who warns us that the digital domain is now undergoing an ideological segregation, where "people restrict themselves to their own points of view—liberals watching and reading mostly or only liberals; moderates, moderates; conservatives, conservatives; Neo-Nazis, Neo-Nazis" and limiting the "unplanned, unanticipated encounters [that are] central to democracy itself" (2001, p. 9).

Granted, we don't live in a color-blind web as envisioned in the early days of the Internet (Daniels, 2012). Race has been architected into our interfaces, as Anna Everett (2008) points out in her book *Learning Race and Ethnicity: Youth and Digital Media*. Race has been coded into the command lines. For example, the early DOS commands designated a 'Master Disk' and 'Slave Disk,' using programming language predicated upon a digitally configured 'master–slave' relationship. Contrary to nascent expectations of 'identity tourism,' where people could escape their racial identity online, much recent evidence has proven the contrary: race relations and communities persists online (Nakamura, 2008).

Undoubtedly, power and privilege are associated with race, playing out in diverse ways in the digital leisure commons. Take, for example, a study illustrated by Watkins (2009), where he draws from a rich collection of 500 surveys and 350 in-depth interviews on the ways youth learn, play, bond, and communicate through virtual leisure geographies. Tracing the movement of young people from MySpace to Facebook, he found a range of language codes used by the youth to differentiate between safe and unsafe people and entire communities. For example, his participants described MySpace as 'uneducated, trashy, ghetto, crowded, and [filled with] predators,' while they described Facebook as 'selective, clean, educated, and trustworthy' (p. 80, p. 83). This movement has been compared to 'white flight' (Boyd, 2011).

Undoubtedly this perspective is rooted in valid concerns. Yet recent empirical findings add a more nuanced outlook to this otherwise bleak scenario. Digital leisure environments can also facilitate a deep sense of bonding, group identity, and color consciousness, making people feel less racially isolated. Looking at this through the perspective of the racially marginalized, research finds that Latinos, Indians, and African Americans post on their walls more about racial themes than other demographics, with an aim to foster tighter and closer-knit social ties. These postings enable them as individuals as well as group members (Grasmuck, Martin & Zhao, 2009).

Or take, for instance, the issue of political polarization and the belief that the Internet fosters these tensions. Within the digital commons, do individuals with a 'conservative' political outlook stay with their own kind while 'liberals' stay with theirs? Are these populations becoming more isolated and segregated online? Matthew Gentzkow and Jesse Shapiro (2011) aggregated data from the United States to assess the extent to which news consumption on the Internet is ideologically segregated. The authors confirm that while there are sites that are ideologically extreme, they account for a small share of online consumption. More promisingly, they found that a significant share of consumers get news from multiple outlets, "visitors of extreme conservative sites such as rushlimbaugh.com and glennbeck.com are more likely than a typical online news reader to have visited nytimes.com. Visitors of extreme liberal sites such as thinkprogress.org and moveon.org are more likely than a typical online news reader to have visited foxnews.com" (p. 1802). An important argument to keep in mind here is that we should not assume that people inhabiting certain spaces share the same beliefs. After all, the reverse is also true where people with different political views interpret the same information in very diverse ways (Gentzkow & Shapiro, 2006).

Basically, it does not take much imagination to create a linkage between these discourses on gated communities of segregation, fear, and exclusion and current concerns that new media sites are becoming more exclusive and compartmentalized through people's choices and behaviors online. Indeed, on both fronts, the central preoccupation is with the systematic erosion of spontaneous social encounters and interactions. Such erosion can have serious repercussions on social empathy as we get more architecturally cocooned. We recognize here that architectures are often rooted in the fear of the unknown and can be deeply radicalized and politicized. Fragmentation of communities is often necessary, as people desire security, trust, and intimacy; ideally, this would not threaten the essence of the public leisure domain. That said, the reality is that we live in an age of deep division where few groups dominate the design, structure, and nature of interaction and tightly hold on to their positions as gatekeepers. We can see why there is a moral imperative to address these divisions to examine how the balance of power can become more distributed across this terrain.

GENDERED GARDENS, POSTCOLONIAL
PARKS, AND THE PUBLIC GAZE

The *beguinages* are small green enclosures. From as early as the 13th century, they could be found nestled between houses in cities of the Low Countries. To the passerby, these walled gardens could appear suffocating as their female inhabitants led lives of chastity and service and appeared trapped in an unconventional life. However, in some respects, these enclosures were more liberating than what women of that time enjoyed, thus earning the medieval metaphor of the 'walled garden of paradise':

> They could leave; they made their own rules, without male guidance; they were encouraged to study and read, and they were expected to earn their keep by working, especially in the booming cloth trade. They existed somewhere between the world and the cloister, in a state of autonomy which was highly unusual for medieval women and highly disturbing to medieval men. (The Economist Obituary: Marcella Pattyn 2013, p. 74)

Extending this discussion, the geographer Mandy Morris draws our attention to the intersections of late 19th-century and early 20th-century discourses on gender and nature. In this context, the garden serves as a site for "a critical reading of the bodily regeneration of gendered identities" while also a space for other possibilities (1996, p. 59). She argues that gardens have long served as important arenas within the larger cultural landscape. They have captured the literary and artistic imagination, particularly as a walled enclosure within a public terrain.

> Whether public or private, represented on the ground, in fiction, or in art, gardens are powerful collectors of symbols. Their meanings, never fixed but continually reproduced and transformed, are historically and culturally specific. (Morris, 1996, p. 59)

In fact, such symbolism has been openly preached through biblical themes where the garden represents female closure, inaccessibility, and solidarity. Undoubtedly then, the walled garden is a powerful metaphor of gendering that can serve as entrapment but can also be reorganized as self-determination within a larger patrician terrain.

   In her beautifully written work *Feminism and the Public–Private Distinction* (1992), law professor Ruth Gavison capitalizes on the historical connectedness of walled gardens to the construction of gender, lending a provocative perspective to this much discussed subject. She argues that by accepting the conventional public versus private dichotomy, we are in fact negating the feminist perspective where "the private is public for those for whom the personal is political. In this sense, for women there is no private, either normatively or empirically" (p. 2). For women, the relegation into the private sphere has historically resulted in marginalization. This feminist narrative needs to be situated within anthropological work across Western borders for a more nuanced and global perspective. For example, from extensive fieldwork with university young women in Riyadh, Saudi Arabia, Leigh Graham (2014) captures a range of digital practices by the girls that challenges these distinct private and public categories. She finds that young girls leverage YouTube space to fulfill a range of desires; they digitally record and upload videos of choreographed fashion shows held in their own bedrooms for instance. Additionally, some girls play out their 'coming of age' through a delightful mix of karaoke, YouTube, and global/local music mix, and some entrepreneurial young women have converted their digital social networks into a commercial space to sell online knockoff *abayas* (large cloaks for women draped from the shoulders or head often covering the whole body). Here, it is evident that YouTube is at once a public leisure sphere as well as a deeply private, personal and significant empowering space, particularly for the cultivation of the female self in Riyadh.

Another important digital platform where we can witness these complex gender dynamics is the massively multiplayer online role-playing games' (MMORPGs') environments. These are particularly significant as the online gaming population is growing at 10 times the rate of the general Internet population and has become one of the prime spaces for leisure online. Furthermore, women are surpassing men as online players, giving us a glimpse of the future community of online gaming. According to Kafai and team's book, *Beyond Barbie and Mortal Kombat: New Perspectives on Gender and Gaming* (2008), even though women's participation in digital gaming have increased significantly in recent decades, gender stereotyping still permeates digital game play. While there is indeed progress—players nowadays can create their own content and somewhat shift dominant structures—the masculinized design of gaming architectures persists. That said, a range of experiments are underway to re-architect these leisure geographies by involving new actors who are more sensitized to diversity in the process of gaming development. The fact remains that within these walled gardens, deep stereotypes pervade.

While much literature has emerged on the sexism that pervades these landscapes, an interesting alternative perspective investigates why women participate and engage in these spaces even as they are aware of this gender discrimination. Themes of social interaction, mastery and status, team participation, and exploration are often evoked when framing female engagement with digital games. Indeed, women seem to enjoy the culture of gaming, where much emphasis is on socializing and building community through chatting, connecting with people, and forming and maintaining relationships. Female participants see these as essential avenues of pleasure. Also, when it comes to violence in gaming, through their avatars, women can experience an activity that is normally demarcated as too aggressive for them and is thereby repressed in social life. This gives them a tremendous sense of power that influences their choice of avatars and gaming tools. This signals that game architects do not necessarily need to swing in the opposite direction and make a game stereotypically 'female' or 'pink' in design, since women are already participating in a nonconventional narrative not always available to them in social life.

> The traditional approach of 'pink games' presents its own set of challenges . . . games that simply focus on friendship and sociality may overlook the fact that 'girls are looking for games which also push them to take risks and where there is a chance to be absolutely and unequivocally dominant' and further suggest that there may be unintended consequences to gender-specific software: girls may be less likely to benefit from developments in the gaming mainstream if they believe that only 'girl games' are appropriate for them. (Taylor, 2003, p. 46)

Basically, when we look at the range of ways in which women are moving through these gaming platforms and "how power and privilege are located in the moments of interaction" (Gajjala, Rybas & Altman, 2008, p. 1131), we can gain much insight into the changing landscape of womanhood. While females continue to be subjected to these gardens walled by masculinity, this by no means equates to complete subjugation.

Therefore, by applying the gender lens, we unravel a rich texture to garden politics that offers a more inclusive perspective on leisure geographies, virtual and material. In this spirit, let us exercise another lens on these terrains: the lens of postcolonialism. Judith Roberts (1998) describes the English walled gardens in India during the colonial era as the re-creation of the British landscape within an alien and hostile territory. Walled gardens became essential to sustain one's 'Englishness' and protect from foreign influences. Much correspondence has been found from that time period that documented how there was a constant feeling of being observed by the natives, creating a feeling of being walled in: "I have the feeling I am perpetually being watched as I dare say I am," said one of the British wives (in Roberts, 1998, p. 116). This illustrates the notion of reverting the 'colonial gaze,' as Edward Said, the renowned public intellectual and author of *Orientalism*, explains: "Orientals were rarely seen or looked at; they were seen through, analyzed not as citizens, or even people, but as problems to be solved or confined" (Said, 1978, p. 207). What was also interesting was that the British walled garden was not re-created in exactitude but, instead, a new breed of park space emerged that blended the native with the foreign culture:

> The lawn was central to the maintenance of the fiction and contributed to the feel of Englishness; but the garden was, in fact, something very different. The British gardening community in India had gradually

evolved a distinctive cultural type; it was not an 'English' but an 'Anglo-Indian' garden. (Roberts, 1998, p. 134)

Similarly, social network platforms introduced to postcolonial countries by the West can be seen as part of the grand mission of extending their power and reach to an emerging market populace. In the name of mitigating the digital divide and setting a path to prosperity and democracy, these digital spaces can be perceived as a neocolonialism of kind. Taking this perspective, however, would be akin to undermining a complex and multi-channeled flow of power. It is more appropriate to view new technology space as negotiated, where often the gazer is also the gazed upon. More concretely, in her book *The New Argonauts: Regional Advantage in a Global Economy* (2006), AnnaLee Saxenian argues that there has been tremendous focus on Silicon Valley in California as an oasis of digital innovation. This eclipses the basic reality that many of the 'architects' of these innovations are foreign-born, technically skilled entrepreneurs who travel back and forth between Silicon Valley and their home countries. Furthermore, major outsourcing hubs that serve as critical back offices for these power centers are located in places like Bengaluru in India, now termed the 'Silicon City.'

These profound transformations and global shifts in the economic landscape also suggest distributed power. For instance, Indians are constantly 'gazing' at the digital movements and inhabitations of people online, particularly those in the global North. Through their coding, they are shifting online activities and behaviors within these privileged territories. While serving as 'gardeners' of these spaces, they are also active gazers. In other words, Saxenian remarks, "the architects of new technology spaces may appear to emerge from the West but in fact cannot be disassociated from other cultures and nationalities and in fact, due to their unique exposure to a range of markets, such argonauts become harbingers of change, exposing the West to new tastes and making exotic behavior the norm" (2006, p. 276). Another way of conceptually framing this, which applies to both the supposed colonial and virtual gardens, is the growing notion of 'sous-veillance,' or what can be viewed as distributed, diffused, and bottom-up surveillance. Indeed, this is an age where our mundane acts within social leisure networks serve as data to be mined by corporations and the state. That said, we are also exercising "watchful vigilance from underneath" (Bakir, 2010, p. 16) by monitoring companies and the state on Twitter and Facebook and critiquing their practices, policies, and politics.

To summarize, feminism and postcolonialism re-territorialize our perspective. They do so in a way that demands our questioning of the normative ways in which we view walls as we inhabit our leisure terrain, both virtual and real. The position of the gardener, the patron, the architect, and the stroller are in flux. We are often inundated with popular stereotypes on what, for instance, women and people from the global South need and want, and both are often wrapped in discourses of victimhood. We have to expand our perspective of walled gardens in contexts commonly seen as patriarchal, masculine, Western, and imperialistic. By doing do, we acknowledge the agency of the marginalized. By understanding that gaze is multidirectional and situational, we can gain a better grasp of the public leisure environments we live in.

PLAYGROUNDS AND PUBLIC SOCIALIZATION OF THE YOUTH

The years 1880 to 1920 witnessed a major shift in the vision of child-rearing spaces by urban social reformers. Not coincidentally, progressive pedagogy had become popular during that time under the likes of tireless proponents such as John Dewey, the liberal educationist. He was instrumental in pushing for a more democratic and public involvement in the process of child development. So it was not a surprise that public demand and government agendas led to a concerted effort to extend children's play spaces from the private confines of the home into the public (Cavallo, 1981). This signaled that it was not just the responsibility of parents to socialize children, but also the state (with the school system by its side). This new movement propelled the government to take on the role of a 'parent,' concerning themselves with issues that were once personal and private. There was rigorous discussion about the design and purpose of playgrounds in the making of the ideal citizen, the new generation of hope. In justifying investment for this new semi-public leisure development for the youth, a local politician of that time, Frederick H. Gillett of Massachusetts, argued that "we cannot afford to be niggardly in anything which promotes the healthy development of children on whom depends the whole future of the nation" (in Cavallo, 1981, p.25). While most reformers agreed that supervised play within these walled gardens was a modern necessity for the socialization of the child, there was constant discussion on the extent and nature of such supervision. Some officials argued that it was essential to keep a close watch on the interactions of the youngsters to protect them from antisocial amusement activities and steer them towards acceptable forms of play. What constituted as 'acceptable' was contentious as there continued to be some distrust on the impact of free leisure and play on the work ethic of the youth. Others saw this view as too paternalistic and socialistic. One of the local politicians pointed out, "If you take this step now, and coddle and supervise him and direct and manage him when he is on the playground, where is he to learn the lesson of self-reliance that makes for sturdy Americanism and self-reliant American citizenship?" (Harrison, 2004, p. 136)

Even today these discussions are heard, as several walled gardens have emerged online. These spaces are funded by the state and constructed in partnership with several educational agencies to serve as modern playgrounds for the youth. They are designed explicitly to entertain as well as educate, or basically 'edutain' the youth and model them to be active and responsible e-citizens. Experts mainly populate the content within these boundaries and the relevant authorities supervise these platforms. Currently, much debate centers on the extent to which these architectures should be standardized and controlled by the authorities, negating the over-arching fact that the youth today are part of a digital participatory and open culture (Dron, 2006). In terms of work and play elements within these platforms, we see their focus tilt towards the more laborious, making them resemble labor camps more than playgrounds. The fact is, when the state gets involved in the design and control of the software of these walled gardens, there is a tendency to apply Fordist efficiency and social morality to these platforms. This often translates to deeply inflexible and utilitarian-biased infrastructures that moderate the movement of the youth at every step. Interestingly, there are some alternative platforms that are opening up through hypermedia systems, connecting to a more open and larger digital leisure environment. In this age of the TED cult, several edutainment initiatives have occupied the public terrain of YouTube and other Web 2.0 spaces, blurring the architectures of social network sites and learning platforms. This movement, albeit still marginal, is challenging these conventional walled gardens for the youth by experimenting with a range of sociocultural and technical affordances that new media platforms provide. However, mainstream virtual playgrounds continue with their reservations: they view Web 2.0 as a threat to the security of the youth, a jungle that is yet to be cultivated.

In fact, public concern over the safety of youth has dominated as leisure architectures loosen their privacy settings. Numerous studies have documented the victimization of youth due to weak regulation of such digital environments. The walls are seen to crumble against perverse publics and corporate vultures that feed on the vulnerabilities of the youth. This gives tremendous impetus to the state and other authorities to protect youth by creating more walled gardens. The impact of these measures has been recently captured in a large-scale survey of trends in youth victimization over the period of 2000 to 2010 (Jones, Mitchell & Finkelhor, 2012). Contrary to media coverage that emphasizes the increase in digital predatory activity towards the youth, it was found that unwanted sexual solicitations declined from 19 percent in 2000 to 13 percent in 2005, and finally to 9 percent in 2010. There was a total 50 percent decline in such reports between 2000 and 2010. In addition, there was a decline in youth reports of unwanted exposure to pornography from 34 percent to 23 percent between 2005 and 2010. Disturbingly, however, online harassment seemed to have increased from 9 percent in 2005 to 11 percent in 2010 (p. 182). These harassments seemed to be particularly severe among certain demographics; the rates of online harassment increased 50 percent for girls, from 10 percent in 2005 to 15 percent in 2010. What is particularly astounding about this trend is that unlike sexual solicitation that came from the outside, harassment is emerging from within the walled gardens of the youth. Part of the explanation for this is as follows:

> First, more of the harassment may come from within the youth's chosen social network, for example, classmates who have been accepted as friends on social networking sites. Second, mobilization and education against online harassment are not as longstanding and intensive as that against sexual solicitation. Public concern over "cyberbullying" only took off in recent years. Now that cyberbullying has become a more widespread topic of news and education, it will be interesting to see whether harassment declines as sexual solicitation has done. (Jones, Mitchell & Finkelhor, 2012, p. 184)

The fact remains that as pressure mounts from parents, educators, and other interested parties, monitoring of digital playgrounds has become the norm. This protectiveness has permeated deeply into all aspects of these infrastructures, leading to certain negative repercussions in the socialization of youth. Karen Malone, a professor of education, calls this new breed of youth the "bubble-wrap generation" (2007). She argues that society is becoming more inward looking with this shift from open play spaces to semi-private spaces for children's play.

> The influence and impact of the world 'out there' beyond the garden, therefore needs to be controlled and managed by parents to ensure it does not seduce the child and expose them to risks. (p. 515)

Here, the architecture of fear fosters a culture of fear within the youth. While no doubt these spaces protect children from unwanted and undesirable advances from strangers, they come at the cost of restricting opportunities to "engage in free play" (Malone, 2007, p. 513). Similar to the argument made earlier on, this approach dissuades spontaneous interaction with those outside of our social capital and keep us cocooned within our own small worlds. Independent mobility of children is also hampered as they are socialized to follow particular avenues, decreasing fortuitous learning encounters and imaginative choice-making. Malone argues that as gates are drawn around play spaces and made into semi-private domains, it instills a broad pattern of risk-averse behavior and inward-looking sensibility. This "climate of fear" has meant that "many parents are restricting children's movements to such an extent these children will not have the social, psychological, cultural or environmental knowledge and skills to be able to negotiate freely in the environment" (p. 514).

Fascinatingly, if we look at popular metaphors that frame children's spaces, Friedrich Fröbel, the German pedagogue and one of the key figures of modern education, came up with the term 'kindergarten,' which translates as the 'children's garden.' Across Europe and the United States in the late 19th and early 20th century, the rise of liberal child-centric pedagogy came with a metaphorical package of viewing children as 'unfolding plants' in the garden of learning. An influential British educationalist, Margaret McMillan, conceptualized the future of schools as 'a garden city of children.' While indeed children are freer to play within these walled gardens as they are watched over, there is the constant temptation to micromanage this generation. In the end, no serious scholarship disputes the fact that children need protecting. The question however, is about gauging the extent to which this should prevail and what is traded away for this security.

In today's digital playground, another fear pervades—the loss of privacy. Decades ago, Dan Hunter (1985) predicted that "our revolution will not be in gathering data—don't look for TV cameras in your bedroom—but in analyzing the information that is already willingly shared" (p. 32). Indeed, this was an astute remark: we have become complicit in the erosion of our own privacy as we voluntarily make our private lives public on digital leisure networks. This is particularly concerning regarding the youth. As new communication behaviors emerge among teenagers where they explore, construct, and play with their identities on digital playgrounds such as Facebook, measures to protect these youth from exploitation become of paramount concern. Disturbingly, defenses are down within this leisure terrain as teens view this realm as a private recreational space. Hence, they swap details of their personal lives, considering this to be their safe haven. Danah Boyd (2007, p. 4) terms these spaces as "networked publics" and compares them to "unmediated publics like parks," making the point that youth use these sites in a freer and more playful manner to socialize and build a culture. The unique design and structural features of these platforms mimic privacy and create the perception of safe play.

> Asking for this type of information and setting up requirements for membership tend to make kids think it is safe to reveal personal information online. Facebook requires an affiliation with a college or high school, which also creates the idea of a semi-private space. (Barnes, 2006, p. 8)

Given the extraordinary time the youth spend within these spaces, they are more susceptible to believing that these realms are more private than they actually are; "those engaged exclusively in recreational domains probably feel this illusion most strongly" (Barnes, 2006, p. 1).

As with any newly designed and inhabited space, there is a learning curve in the understanding of its architectures. Often, it is a matter of time for the inhabitants to learn how to circumvent and play with the barriers to best suit their needs. As a recent study by Boyd and Marwick (2011) points out, the youth are no different. This study disputes the narrative of youth naïveté on issues of privacy on social network sites. Using a large-scale survey on attitudes and practices on Facebook of 18- and 19-year-olds in the United States in 2009 and 2010, they found that teens have developed sophisticated sociolinguistic strategies that enable them to carve out privacy within Facebook's open gardens. By communicating through personally manufactured codes that are understood mainly by their peers, they are able to be public and private at the same time. The authors emphasize the contextual nature of youth practices, wherein "networked publics are shaped by their interpretation of the social situation, their attitudes towards privacy and publicity, and their ability to navigate the technological and social environment" (p. 1).

To conclude, multiple institutions participate in the socializing of the youth, from the most private unit of the family to the most public unit of the state. All parties are in consensus over the fact that youth require protecting and nurturing. Yet interests evidently diverge when in negotiating the architecting of these digital playgrounds. Society is in constant flux about the trade-offs: it tries to create an environment that nurtures independence of spirit while continuing to watch the youth with a paternal gaze. Educators are forming new and closed social network sites as 'e-safety gardens.' They believe that, especially for children, there is a need for deeply curated and mediated spaces that are less at the mercy of corporate agendas and perverse individuals. On the other hand, highly regulated playground spaces are seen to turn youth into an inward-looking demographic, driven by the element of fear. Through this design, the outside world is presented as a dark, dangerous, and high-risk place.

Youth as a demographic unit are hardly homogeneous. Different segments, based on sex, race, ethnicity, and nationality, require more protection than others; dominant social norms do bias playground architectures. While parents, educators, and the government strive to provide secure spaces for youth to freely explore without manipulation, we need to be careful of making these spaces too sanitized, mediated, and regulated in their architectures.

COMMUNITY GARDENS AND JOINT GOVERNANCE

To view the future of urban parks as walled gardens is to accept the limitations and predictability of the social imagination. When we do so, we are resigning ourselves to prescribed park terrains that are built based on the need for control, regulation, and mediation. Distrust is an embedded value within such social architectures. When we pry into the historic countercultural movement of urban green spaces, we are thankfully relieved of such notions. Community gardens are one such response and statement to the market-led structuring of public leisure environments. The simple act of squatting has become a powerful anarchic technique around the world to reclaim public right to private property. Far from serving as peripheral politics of the land, this occupation and communal participation is central to the history of public space, including that of urban parks. The author of *Shadow Cities*, Robert Neuwirth (2005), shares a powerful statistic: there are one billion squatters globally, that is, about one in every seven people on the planet. The squatter movement gained strong momentum against propertied interests, laying out an alternative vision for the public sphere. In her work on community gardens, Karen Schmelzkopf (2002) remarked:

> Community gardens are one of our most participatory local civic institutions . . . There has been wide recognition of the worth of the 'sweat equity' labor of the volunteer gardeners. Finally, there are the benefits of open space and nature. (p. 332)

Similarly, people are far from abdicating social media space to private and commercial control. The 'commons' is frequently invoked in discussions of the Internet, often as a reminder and warning that something must collectively be done if we are to protect Web 2.0 space from strong corporate and state forces at play:

> The insidious combination of capitalist ownership at the infrastructural layer and corporate control at the content layer portends an impoverished future for the Internet: instead of a virtual commons, cyberspace will increasingly resemble private property, hemmed in and protected by state and market forces. (Lessig, 1999, p. 14)

As explained in the sections above, the private sector has indeed gained tremendous ground in usurping and transforming the character of our common digital space. While the gating of public leisure space may appear inevitable, there are active forces resisting this and revitalizing the public dimension. This collective performance to ignite and sustain the public sphere can be viewed as an 'information commons' enabled by open source, open access, fair use, and neutral networks (Aufderheide, 2002). The value of such spheres is entrenched in the right to access free software and community participation in institutional knowledge and policy making. After all, the 'participatory culture' of Web 2.0, while admittedly harnessed by corporations, also serves as vital grounds to foster public good.

Social collectivities are in the midst of generating alternatives in digital sociality. One example is Diaspora, a social networking platform developed as an open source alternative to Facebook, which defines itself as "privacy-aware, personally controlled, do-it-all, open source." The platform is sustained via public donations instead of advertising. It is argued that the key to have open public space without overt agendas or mediators is to be free of market constraints (Benkler, 2006). Clearly there is some desire to go back to the original ethos and mission for governance of the Internet, where commercial interests did not dictate and control. The open source movement is one such means to sustain the culture of openness through self-organized, autonomous, decentralized, non-hierarchical, and volunteeristic processes, where anyone can develop code and contribute to this project. Ironically, while such a movement has created one of the largest and most successful operating systems, GNU/Linux, it has also spurred a major and profitable industry worth millions of dollars. A similar fate has been bestowed on the web browser Firefox. While it is open source, it has become Microsoft's competitor and is used by several major multinational corporations to further their profit. Furthermore, power enclaves of expertise are being created within the classic online commons open source model Wikipedia, begging the question of whether it is possible to maintain the non-hierarchical ideals of open public social space (König, 2013). These realities may surprise us given the romanticism around the open source software and other collective movements to 'free' the Internet.

To address this more coherently, let us shift our attention back to the field of urban planning and the governance of urban green space. The human geographer Marit Rosol (2010) points out that community gardening is not always the antithesis of market forces. Rather, it is a materialization of a complex shift in relations among state, non-state, and corporate actors. These should not be viewed as simply opposite and reactive measures to market mechanisms. Indeed, on the one hand, these spaces are grassroots in origin and belief. On the other hand, they are also entrepreneurial manifestations that have risen because of decreased responsibility by the state and an increasing neoliberal approach within urban governance. Such an approach is based on outsourcing state services. This perspective builds on critical work, such as that of Ash Amin (2005), who argues that civic participation in the greening of social space is compensation and expression for the state's retreat from its welfarist functions. One of the consequences could be the alignment of non-state actors and their collective action with that of the state and market. In her book, *Avant Gardening: Ecological Struggle in the City and the World* (1999), Sarah Ferguson illustrates this through the story of grassroots greening of New York and how collective participation benefited both the private sector and the state. Indeed, public participation in 1977 was able to take on 27,000 vacant lots littered with trash and rats, used mainly for prostitution and drugs, and convert them into more habitable spaces within the city:

> Liz Christy and a band of like-minded activists called the Green Guer-
> rillas began taking over abandoned lots on Manhattan's Lower East
> Side . . ."You could not have picked a more unlikely place to start a

garden," recalls Bill Brunson, an early Guerrilla . . ."To put a garden there—in what was probably the ultimate slime spot in the city—that was unheard of." (Ferguson, 1999, p. 64)

Across the United States, these actions transformed the urban landscape and served as an inspiration to others. They also served as a gift to the real estate magnates by increasing property values in the Lower East Side, and were able to purge the unwanted public, such as the hustlers, panhandlers, and prostitutes. Such a task would have been more politically delicate if executed by the state. Thereby, civic participation can be deeply beneficial as it has "the potential to depoliticize urban governance practices and effectively discipline community organizations into forms of participation that is more manageable for the state" (Elwood, 2002, p. 123). This co-optation of citizens and their meaningful practices for the state and private agendas indeed resonates within the digital commons.

So is joint governance an ineffective and unrealistic means of sustaining a public leisure space? Is communal-driven space too utopic for sustaining equitable social and leisure activity? If we are to look at the material dimension, the tragedy of the commons is a well-known concept: it is argued that to sustain public social space, there is a need to create boundaries of exclusivity. The popular belief here is that such enclosures of the commons through private property are the only viable means of protecting finite resources. However, in later studies, arguments were made against such determinism. For example, Ostrom (1990) found that communities create a variety of social norms to sanction actions that go against the common interest of the public. In fact, Bollier (2002) advocates communal governance of public space as "social infrastructures" and argues it is more effective against abuse of such space; "trust, reciprocity, a history of shared commitment and a robust community can overcome many of the alleged failures of the commons" (p. 6). Lessig (1999) agrees that managing shared resources by communities is possible and, as popularly believed, the tragedy of the commons is not an eventuality. Partly, we need to clearly define notions of collective ownership of public beneficial spaces like urban parks; the operationalization lies in the details. While privatization is not a natural end to this game, scholars admit that it requires sustained momentum to keep social production free from corporate interest:

The basic characteristics of a commons regime are cooperation, decentralization, communal assets, and a public service ethic . . . The tragedy of the commons is therefore a misnomer: if an individual overexploits a common resource based on self-interest, that is privatization, that is enclosure, and that is quite the opposite of a commons regime. The tragedy of the commons is the tragedy of enclosure. (Anderson, 2008, p. 29)

At times, more participation can be better for instilling social value and meaning to public leisure space. Further, regulation of such space needs to be done by groups instead of individuals. Although hierarchies may be formed, it is not necessarily antithetical to the public nature of the domain. In other words, it takes private initiative among social members of these spaces to sustain their public character; "the difference between exercising a right and participating in the definition of future rights to be exercised . . . [that] makes collective-choice rights so powerful" (Ostrom, 1990, p. 59). In fact, Ostrom proposes a blueprint for governing the commons through a set of five design principles, namely: clearly defined boundaries, the design and enforcement of rules, reciprocity (the equal exchange of goods and knowledge), building trust and social capital, and communication channels.

Overall, we can say that while indeed it is a struggle to create and maintain an alternative to corporate public leisure spaces through communal ownership and governance, it is not an eventuality. In fact, it is argued that a certain amount of private initiative and hierarchy can be necessary to sustain the public nature of such spaces. This is possible as long as these spaces are not tied solely to corporate ends with purely profit-focused interests.

To conclude, the ideal and rhetoric of the 19th-century public park as a communal and open space have carried over to this day, especially in the virtual sphere. Environments of leisure, both offline and online, both for the public and by the public, are seen as essential to foster a community that values diversity and freedom of practice from private-sector mediations. However, as this chapter reveals, there are deeply troubling trends focused on the gating of public space as semi-private 'walled gardens.' These walled gardens are bound by corporate interests and state agendas. Fear and distrust of the unknown are embedded in these network architectures, threatening to shape future inclusive publics. This ideology can filter down to more micro-levels of exclusion, separation and division, creating a designed public that is averse to unmediated public leisure. This chapter argues that such exclusion is fundamentally antithetical to an egalitarian and outward-looking society. Architectures are value driven; hence we need to pay special attention to the moralities dictating their design and governance. We need to capitalize on existing knowledge regarding material leisure spaces. These spaces include a range of walled garden manifestations: gendered gardens, postcolonial parks, security parks, and community gardens. All of these serve to illustrate and highlight contemporary concerns influencing web architectures of public sociality.

We learn that the globalizing of such spaces can homogenize these terrains. Governing these spaces requires not just legal efforts but routinely performed micro-governance, enabling and normalizing practice. We discover that architectural cocooning can come at the price of civic responsibility and the breeding of a new generation of risk-averse youth. Further, we see that scaling of architectures, both virtual and real, simulate an illusion of sufficiency and diversity of choice. These enclosures grow as corporations shape architectures to limit the movement of participants from outside their marked boundaries. In the name of efficiency, predictability is fostered at the price of spontaneity and openness to the unexpected. This has a particularly negative impact on youth. Corporate spaces capitalize on youth vulnerability in order to portray the space as the norm for their play. This corporate watchful gaze impacts the nature and quality of their interactions. To conclude, this chapter emphasizes that this is not a deterministic outlook. Rather, it is a means to create urgency in fostering alternative web architectures that value openness and non-corporatized mediated action.

5 Corporate Parks
Usurping Leisure
Terrains for Digital Labor

> The notions of urban 'garden districts' and 'green belts' were expressions of the Humanist desire to build a better world in which leisure and work reinforced one another in a spirit of harmony.
>
> Chris Rojek, *The Labour of Leisure*

> Today, communication is a mode of social production facilitated by new capitalist imperatives and it has become increasingly difficult to distinguish between play, consumption and production, life and work, labor and non-labor.
>
> Trebor Scholz, *Digital Labor: The Internet as Playground and Factory*

Work defines us. Play liberates us from these definitions. This conventional wisdom is continuously readdressed as our meaning for both work and play transforms over time. Every society is driven by a vision of social life and, intrinsic to this template, lays the intertwined architectures of labor and leisure. For instance, workspaces have undergone a tremendous change as employers have evolved in their understanding of what counts as productivity. In this innovation obsessed economy, the common wisdom among many companies is that to attract the best talent, a new corporate culture is needed, sensitized to the workers' larger well-being. Some companies are focusing on the very space within which such talent can be nurtured—the 'office.' The typical gray cubicle infrastructure is making way for a different work environment. Pool tables, volleyball courts, video game parlors, pianos, Ping-Pong tables, and yoga stations are becoming a signature of these new labor landscapes (Kjerulf, 2007). Bicycles, scooters, and slides enable employee movement. Play is infused in the design and shaping of the reception area and boardrooms. The individual company gives way to an ecology of corporations situated in park spaces, resembling a university campus (Daskalaki, Starab & Imasa, 2008). From the cubicle to the hammock, there appears to be a shift in perception among some key companies on what counts as a productive space in today's business market. It isn't surprising that creative and technology industries like Pixar, Apple, and Google have embraced the re-architecting of their corporate settings to resemble play spaces (Chang, 2006). Innovation is their business: It is believed that the less regulating, confining, and spatially predictable a work environment is, the more likely it will be to generate new ideas and enhance performance. These corporate parks share more with the ethos of public parks than with the signature office, simulating a place that is relatively free from typical business routines and least marked by institutional practice.

These new labor geographies are not confined to the West. Emerging markets have risen dramatically in the last decade and are being viewed less as back offices and blue-collar workspaces and more as drivers of today's global economy (Vaidyanathan, 2008). They are entering the world of business innovation, luring an increasingly global, cross-cultural, and diasporic employee base. We see this manifested in ambitious new corporate designs of workspaces, be it the Infosys campus in Mysore, India, that serves as a green oasis for its employees or the impressive Shanghai Huawei Technologies Corporate Campus embedded in a lush wetland landscape, representing a corporate culture that seeks to connect work with nature. Most of these companies are part of a larger industrial, science, or technology park where the concentration of expertise and knowledge is being promoted for regional development (Goldstein & Luger, 1990). This appropriation of leisure space is not endemic to these niche industries. It is becoming more commonplace among diverse private-sector entities, including the healthcare record industries. Epic Systems is a good example: situated across 800 acres of former farmland near Madison in Wisconsin, it supplies electronic records for large healthcare providers, such as the Cedars-Sinai Medical Center in Los Angeles, the Cleveland Clinic, and Johns Hopkins Medicine in Baltimore. On entering this work environment, one is struck by the eccentric and playful atmosphere created through the choice of architecture.

> The push to move the nation from paper to electronic health records is serious business. That's why a first look at the campus of Epic Systems comes as something of a jolt. A treehouse for meetings? A two-story spiral slide just for fun? What's that big statue of the Cat in the Hat doing here? Don't let these elements of whimsy fool you. (Freudenheim, 2012, p.2)

The incentive to design the corporate office in the manner of a play space was to "draw programmers who might otherwise take jobs at Google, Microsoft or Facebook" (Freudenheim, 2012, p. 3). Hence, we come to an understanding of how productivity, creativity, and innovation are tied to the spatial context within which they arise.

Jason Fried, an emerging technology guru and coauthor of *Rework* (2010), pontificates about new ways to conceptualize working and creating. He attacks the sacred domain of the office, underlying its irrelevance

Figure 5.1   Infosys Headquarters in Bengaluru, India.

in a society that hungers for creative capital. In the initial decade of social media, corporations panicked (and still do to a large extent) about the infiltration of leisure into work. Corporate reaction manifested in litigious reactions to these micro-deviances. Fried believes that corporations are deeply misguided, "Facebook and Twitter aren't the real problems in the office. The real problems are what I like to call the M&Ms, the Managers and the Meetings" (Fried, 2010, p. 1). Businesses are missing the point, he feels. Their focus should be on the current design of corporate space that often undermines real work through the embedded structure of repetitive business practice. After extensive interviews with numerous professionals, research found that most people got their work done when not at the office.

> Companies spend billions on rent, offices, and office equipment so their employees will have a great place to work. However, when you ask people where they go when they really need to get something done, you'll rarely hear them say it's the office . . . I don't blame people for not wanting to be at the office. I blame the office. (Fried, 2010, p.5)

Figure 5.2   Google campus, Mountain View, California, U.S.
Source: Sebastian Bergmann (originally posted to Flickr as Google Campus) (CC-BY-SA-2.0 [http://creativecommons.org/licenses/by-sa/2.0)]), via Wikimedia Commons.

A decade ago, corporations' instinctive and sole response to social media within the work domain was to sue online business-bashing employees. We were inundated with headlines of companies taking employees to court for posting a negative comment on Facebook or Twitter. Today, the situation is more complicated: Corporations are realizing the implications of these disputes on public relations. They are also becoming aware of the fact that these digital leisure terrains can benefit them if used strategically (Gely & Bierman, 2007). Large corporate giants like Microsoft, once the nemesis of social media, is today an enthusiastic host to more than 1,000 in-house web logs, where workers can offer opinions on everything from astrology to C++ programming. "We see blogging as a great opportunity," says Sanjay Parthasaratby, Microsoft's corporate vice president. "We get greater insight into what is going on with key technologies inside the company" (Edwards, 2004, p.1). Blogging has become so popular at Microsoft that the company offers a web clearinghouse to highlight their various blogs and bloggers. Other companies, such as American Airlines, are using blogs to give employees more channels to management. At IBM, employees from 30 countries use blogs to discuss software development projects and business strategies. Hot Topic, a 690-store retail chain, recently launched an internal social site for employees to share ideas and data.

We see businesses virtually extending their presence on sites conventionally demarcated for online social and leisure purposes, such as Blogger,

Twitter, and Facebook (Hermann, 2006). As seen in the examples above, there is also a tendency to create walled gardens for employees, so companies can closely monitor and architect features that are conducive to their unique settings. This is seen as enabling the restructuring of the top-down private-sector model to a more employee-driven and client-oriented culture. Further, web zones like LinkedIn are seen to facilitate networks and collaborations among employees and companies; the intent is that such connectivity of ideas and people will proliferate new ways of thinking and doing in the business sector (Guerrier & Adib, 2003). Interestingly, LinkedIn's digital space has transformed over the years and resembles more a social and leisure venue now than a utilitarian space that fosters work connections. The rise of digital labor adds a new dimension wherein people across nations work on a project and are paid for their creative input (Howe, 2006). This posits a challenge to the design of workspaces as they cater to a temporal, diverse, and sporadic global labor market. Furthermore, 'gamification' has become a new buzzword within the labor landscapes. By infusing game dynamics into the work culture, this trend is expected to enhance employee engagement and problem solving.

In this chapter a number of parallel pursuits are explored. We look at the historical proliferation of industry, technology, and science parks and their contemporary manifestations. We confront private-sector appropriation of social media spaces, both digitally and materially. Such a focus is essential to gain insight into the role of the leisure commons in fostering productivity, innovation, and creativity in the workplace. This chapter applies the metaphor of 'corporate parks' to examine how business geographies extend to and influence social media spaces as they strive to realign the labor and leisure domain for innovation and employee satisfaction. This trend is historically positioned by examining how leisure space has been legitimized in the past for productivity. We build on the blurring boundaries of work and play, and delve deeper into certain social media trends such as free labor. The foundation is laid to investigate how corporations adopt and usurp common leisure grounds for innovation, strategic networks, and regional development. Overall, this chapter addresses the 21st-century architecting of business spaces, materially and virtually, and how it relates to the changing perspectives on labor, leisure, and innovation in work cultures.

BLURRING THE LABOR AND LEISURE TERRAIN

The Historical Struggle for Leisure in the Labor Landscape

Leisure has traveled quite a way to gain credibility. Puritans lost their grip on the worldview of 'leisure as sin' particularly during the industrial era in the second half of the 18th century. The utilitarian mantra 'an idle mind is a devil's workshop' gave way to the popular proverb 'all work and no play makes Jack a dull boy.' This was a revolutionary shift in human perspective, but it did not arise without a struggle. The reputed historian Roy Rosenzweig, best known for his book *Eight Hours for What We Will* (1985), poignantly captured the uphill battle of the labor movement for an eight-hour workday and the subsequent rise of more urban leisure spaces such as public parks.

> On December 2, 1889, hundreds of trade unionists paraded through the streets of Worcester in a show of strength and determination. 'Eight Hours for Work, Eight Hours for Rest, Eight Hours for What We Will' declared a banner held high by local carpenters . . . the actual quest for 'eight hours for what we will' reverberated through the labor struggles of the late nineteenth and early twentieth centuries. As a compositor who told the U.S. Senate Committee on Relations between Labor and Capital in 1883: 'A workingman wants something besides food and clothes in this country . . . he wants recreation. Why should not a workingman have it as well as other people?' And in industrial communities across America workers fought not only for the right to time and space for leisure but also for control over the time and space in which that leisure was to be enjoyed. (p. 1)

Undoubtedly, the labor movement contributed to this shift in perspective on the labor–leisure relationship and work culture. At the same time, there was recognition from management that productivity at work was enhanced by leisure in social life (Roberts, 2006). Leisure was found to have a legitimate role after all.

Leisure was defined as that which was not work, or that which was in relation to or a product of work. In other words, leisure existed to serve labor or labor existed to produce leisure but the "twain were believed to not meet—leisure and labor are two sides of man's shield; both protect him. Labor enables him to live; leisure makes the good life possible" (Woody, 1957, p. 4). This perspective has its roots far back, as evinced through Aristotle's observation on the relationship between these two domains, stating that "we labor in order to have leisure" (Rosenzweig, 1985, p. 31). Clear dichotomies were laid out in conceptualizing these two realms: Work was a necessity that served utilitarian ends, while leisure was a luxury that was earned through labor. As prosperity grew in the industrial nations, leisure became a more central preoccupation. From the 'labor for the many, leisure for a few' thinking of feudal times, the new phenomenon of the modern era was the massification and democratization of leisure. In his book *Leisure in Contemporary Society* (2006), Kenneth Roberts extrapolated the pervasive perspective of the mid-1900s of spillover and compensation that linked labor and leisure.

> Two main kinds of work-leisure relationship were identified . . . the first was spillover, where work-based relationships, interests, social

and technical skills spread into leisure. A common example given was white-collar workers who were able to use occupational skills in running voluntary associations. A less attractive spillover was said to be the workers with routine, mundane, mind-numbing jobs whose mentalities were so stunted that they were content to spend most of their leisure being passively entertained. The second type of work-leisure relationship was compensation, where individuals used their free time to seek experiences they could not obtain at work. An example frequently offered was the desk bound executive who played sport in the evenings and at weekends. Another was individuals who were denied opportunities to display initiative at work who used their leisure to demonstrate their autonomy ostentatiously. (p. 57)

While this framing appears outdated, particularly in wealthy service economies, it arguably can be applied to the substantive youthful workforce in emerging markets such as China and India. A recent article in *The Economist* ('India's demographic challenge: Wasting time', May 11, 2013) highlights the fact that India itself will soon account for a fifth of the world's largest working-age population, stuck in the industry sector of routinized and mind-numbing jobs. Hence, while the wealthier segment of the world's population struggle with the blurring of leisure and labor through new mobile technologies, for instance, much of the global South continues to be seeped in 19th century notions of labor and leisure.

In some cases such as the United States between 1890 and 1940, Fischer (1994) found that American leisure grew exponentially, even during the Great Depression era of the 1920s and1930s. This is particularly interesting as the popular conception of leisure is correlated to economic security, where higher classes have more access to leisure (Veblen, 1899; Florida, 2002). While no doubt there is evidence to support this perspective, it is still just one part of a larger matrix of leisure–labor relations. Looking across cultures and contexts, one finds that in spite of the lower financial status, poor communities carve out opportunities for leisure to sustain their cultural and social capital (Hutchison, 1988; Marshall et al., 2007; Snir & Harpaz, 2002).

Despite this evidence, we see current policies in the global South reflect an archaic notion of labor–leisure relations among the poor. Numerous economic development plans to create digital access for the poor rest on the premise that this demographic will use new media spaces for utilitarian and work-related ends (Arora, 2012a; Arora & Rangaswamy, 2013). Several schemes have risen for farmers to be able to check crop prices online, for women to access healthcare information, and for the youth to find employment. However, a number of field studies (Rangaswamy & Toyoma, 2006; Ganesh, 2010; Arora, 2010) have revealed that contrary to these expectations, the poor in the global South appear more aligned with the typical user of the global North in their usage of new media. This body of research has found that the poor dominantly use these platforms for socializing, gaming, and consuming popular media and pornography. I have argued in prior texts about this matter, pushing for the examining of leisure within the highly instrumental biased worldview of international development:

> The neoliberal view espouses that the poor will 'leapfrog' conventional and chronic barriers for higher socio-economic mobility. Yet, if equity between the 'Third' and 'First' World is to be achieved, we should expect that the poor, just as the rich, the rural, just as the urban, folk, will use computers for 'frivolous' and 'trivial' purposes. One can argue that this persisting tension stems from a morality of poverty where the pragmatic and ameliorative are the main benchmarks concerning Third World computing. After all, the field of ICT4D emerged and arguably continues to be rooted in postcolonial discourse and practice with a focus on necessities for human and social development. Yet, through this narrowed lens, we can miss the actual engagements and ingenious strategies that the poor employ to cope and escape from their current plight. Entertainment is a key tool here with class taking a backseat. (Arora, 2012a, p. 99)

While the degree of leisure consumption pervades across diverse social categories, there are often distinct differences in the nature of practice. This diversity often stems from the unique social and historical contexts within which these groups reside. Take, for example, the ongoing discussion of why, when it comes to work and leisure, there are distinct differences between the United States and Europe. A report by the National Bureau of Economic Research concludes that a combination of tax systems, labor laws, and other structural mechanisms shape the perspectives towards these two entities:

> Our punch line is that Europeans today work much less than Americans because of the policies of the unions in the seventies, eighties and part of the nineties and because of labor market regulations. Marginal tax rates may have also played a role, especially for women's labor force participation, but our view is that in a hypothetical competitive labor market without unions and with limited regulation, these tax increases would not have affected hours worked as much. Certainly micro evidence on the elasticity of labor supply is inconsistent with a mainly tax based explanation of this phenomenon, even though 'social multiplier effects' may 'help' in this respect. (Alesina, Glaeser & Sacerdote, 2005, p. 30)

Another perspective can be found through the feminist approach to this dichotomy. Here, 'work' in the industrial sense is seen as problematic, negating informal domestic labor and its lack of financial remuneration

(Henderson, 1996). It is argued that women's leisure becomes an invisible act, as it is not tethered to the typical work domain, and that women experience leisure in their own diverse and expansive ways. Contrary to conventional thinking, where the 'modern' replaces the 'traditional' lifestyle and mind-set as per the *Gesellschaft* for *Gemeinschaft* model (Tonnies, 2002 [1887]), it was found that many new leisure behaviors augmented old ways of experiencing leisure rather than uncritically replacing them.

Questions abound: Is leisure becoming more commoditized and commercialized? Is it more a private affair than a public activity? Is it more organized than informal in nature? In some sense, leisure has gained centrality and become an entity in its own right. Perhaps so much so that one can argue that the pendulum has swung to the other side, where leisure has generated much attention (and at times fear) with regard to its role in business spaces and practices (Tapscott, 2009). At the heart of this momentum is the promise (or threat) of new media exponentially scaling and invading established work norms, potentially destroying the boundaries between labor and leisure. Should we be threatened by this blurring of borders?

Constant Busyness: Exploitation or Liberation?

In this mobile telecommunication culture, we witness entanglement as people find ways to incorporate leisure in their work life and work in their leisure time (Du Gay, 1996). Conventional work patterns are being challenged because of new media affordances. Digital platforms allow for the breaking away from the nine-to-five workday and the possibility for part-time and remote workers not indigenous to the company's location (Gershuny, 2005). While these new technology spaces have allowed for the de-anchoring of the employee from the workplace, they have entrenched the worker in a cycle of constant, albeit intermittent, laboring. This has fed into the already growing commuting and outsourcing work cultures and the creation of an 'always on' employee and work ethic. We find ourselves in an era of busyness where a clear demarcation for work and leisure has blurred with the rise of the 'thumb generation,' savvy 'netizens' who are at the constant beck and call of their clientele and supervisors (Buckingham & Willett, 2006). Capitalistic notions of efficiency and productivity are manifest in the BlackBerry generation, paving way for a work culture that is immediate and constant. The addiction to continually checking updates on our mobile devices became so commonplace that in 2006 'crackberry' became the winner of the 2006 Word-of-the-Year contest by Webster's Dictionary. While we can recognize that the pace of social life has changed because of new media technologies, some claim it is for the better and some for the worse.

As the middle class expands, as choices increase, as mobility and access widens through new technologies, expectations of the type of labor people are willing to engage in have begun to shift. Emphasis is placed on being 'authentic' to oneself by creating coherence between our work and leisure lives. In this perceived individualistic age, "people are encouraged to 'know themselves,' 'be themselves' and 'be true to themselves' especially through their leisure activities" (Guerrier & Adib, 2003, p. 1401). One can argue that the ideal 'job' is constructed around its proximity to leisure and social and intellectual enrichment, stimulating personal satisfaction. Certain corporations are now seeing the benefits of leisure to enhance innovation and creativity at the workplace and are shaping their corporate geographies to reflect this newly embraced work culture. The main difference between the industrial and the digital age is in the popular conception of leisure where, in the former, leisure was to supplement labor while, in the latter, leisure is interwoven with labor. As this blurring gets complicated with the infusion of new media technologies, there is an element of choice in demarcating these practices as more leisure- or labor-oriented.

In her blog, apophenia, the social media scholar Danah Boyd recently posted on the dilemma of defining work in a networked world. She reflects on the fact that she has lost all sense of whether she is working or playing. She questions how to make sense of this blurring of boundaries that grows increasingly complex in the digital age. She recognizes that the extent of control the person has on their 'space, place, and time' is indeed a mark of privilege. Narration of her daily routine gives us a glimpse of how her privilege plays out:

> Today, I have my dream job. I'm a researcher who gets to follow my passions, investigate things that make me curious. I manage my own schedule and task list. Some days, I wake up and just read for hours. I write blog posts and books, travel, meet people, and give talks. I ask people about their lives and observe their practices. I think for a living. And I'm paid ridiculously well to be thoughtful, creative, and provocative. I am doing something related to my profession 80–100 hours per week, but I love 80% of those hours. I can schedule doctor's appointments midday, but I also wake up in the middle of the night with ideas and end up writing while normal people sleep. Every aspect of my life blurs. I can never tell whether or not a dinner counts as "work" or "play" when the conversation moves between analyzing the gender performance of *Game of Thrones* and discussing the technical model of Hadoop. And since I spend most of my days in front of my computer or on my phone, it's often hard to distinguish between labor and procrastination. I can delude myself into believing that keeping up with the *New York Times* has professional consequences but even I cannot justify my determination to conquer Betaworks' new Dots game (shouldn't testing new apps count for something??). Of course, who can tell if my furrowed brow and intense focus on my device is work-focused or not. Heck, I can't tell half the time. (Boyd, 2013, p. 3)

It is worth keeping in mind that this association of busyness with the privileged domain was not always so. If we go back to the 19th century 'man of leisure,' the signal of the wealthy class was to be able to have as much time at one's disposal. The classic text by Thorstein Veblen, *The Theory of the Leisure Class* (1899), was well ahead of its time, underlying the critical point that leisure was in fact a social construction and that because of the specificities of the context of the time, it served as an important signifier of upper social status.

Ironically, today busyness seems to have transplanted leisure, serving a similar purpose. In present-day contexts where work time has reduced, the feeling of busyness seems to be all pervasive as people report being 'on' all the time. Jonathan Gershuny (2005) from the Center of Time Use Research tackles this paradox of busyness by emphasizing the affective in this process. In other words, there has been a remarkable shift in the social value of constantly working, often resulting in the feeling of busyness. There is an implicit aspiration to be immersed in constant laboring, equating this to higher social status. Gershuny reminds us that this feeling can just as well apply to intense and constant leisure activity.

As we shift gears to the context of emerging markets, my own fieldwork in rural Central Himalayas has documented numerous events where busyness is thrust upon citizens, cutting across class sectors. Busyness, akin to a social virus, is found everywhere, even in the rural domain. There is a disjuncture in the conventional relationship between class and busyness. While there appears to be a drive to capitalize on time, it seems to be for a host of practices, including that of leisure. People in rural cybercafés are immersed in the ritual of busyness, and, within this sphere, practices of dating, socializing, and creative play surface. Rather than labor, one frequently encounters leisure activity. In resource-poor contexts, much time accrues because of systemic breakdowns of socio-technical infrastructures. There is a dominant expectation that people will capitalize on their time more productively. Yet what was witnessed was that people found ways to occupy themselves simultaneously with the mundane and the creative, for labor as well as for leisure.

New technology was meant to liberate us from work. And it did. It also freed us up for more labor (Levine, 2005). There seems to be a circular pattern of effectively managing our complex lives with the aid of new technologies. With that comes an acceleration of lifestyle with little room to pause and ponder. The sociologist and feminist Judy Wajcman (2008) argues against the popular discourse of time-space compression and expresses deep skepticism about this determined pace of social life in this supposed postmodern society. Much like the 'industrial revolution' that promised a 'leisure revolution,' new media technologies seem to threaten us with an age of acceleration and scarcity of time and leisure. Instead, she suggests, we need to look at the quality of leisure and how that differs among diverse social groups and contexts. Such a focus creates a more nuanced and rich discussion. For instance, viewing along the gender lines:

the quality of leisure differs in two important respects . . . 'pure' and 'interrupted' leisure, we show that men enjoy more leisure time that is uninterrupted (that is, unaccompanied by a second activity). Women's leisure, by contrast, tends to be conducted more in the presence of children and subject to punctuation by activities of unpaid work. In addition, the average maximum duration of episodes (blocks of time) of pure leisure is longer for men (that is, that women's leisure is more fragmented into periods of shorter duration than men's). It seems reasonable to assume then that women's leisure time may be less restorative than men's. (Wajcman, 2008, p. 66)

She leaves us with advice on how to approach this subject, namely, that we should stay away from a deterministic perspective; rather, focus on how people appropriate digital platforms to take control and create a labor–leisure balance that is suited to their contexts and needs. While no doubt there is a new media focus to this book, the spatial, historical, and sociocultural angle plays a central role in this investigation. Here we should ask: To what extent is busyness associated with ordinary, routine action? Is busyness primarily an outcome and state of work? Does busyness primarily lend itself to fragmentation of sociality? The bottom line being, can technologically induced busyness serve as a platform for sociality, cultural, and creative activity? As we shift contexts, we gain a wider and more complex understanding of the role new media plays in busyness. Also, we need to consider how busyness weaves into the staid dichotomy of leisure and labor, juxtaposing the global North with that of the global South.

WORK CULTURES AND 'PLAYBOR' GEOGRAPHIES

Hobby Farming and Free Labor

Hard labor can be quite pleasurable. Certain leisure domains demand a concerted and continued effort for sustenance. The cultural geographer Hayden Lorimer (2005) brings attention to the historical practice of 'hobby farming' and how it was seen as distinctly different from agriculture. While the latter was clearly regarded as tedious labor, the former was immersed in the ideal of green space, romanticizing the act of cultivation as a deeply pleasurable, collectivistic, and authentic act. In other words, labor here was associated with the "passionate, intimate and material relationships with the soil, and the grass, plants and trees" (p. 86). While the effort could be the same in both domains, the element of choice converted tedium into euphoria. Veblen's conceptual framing of 'exploit' as a form of play is useful in capturing the essence of difference between these two types of laboring. While work in the context of hobby farming is not by any means pure play, it does indicate that embracing challenges and taking on labor wholeheartedly underline the passage of exploit becoming enjoyable. Therefore, the motive that drove these efforts for the hobby farmer was "other than profit, such as pleasure or amenity, and he [*sic*] is not, therefore, dependent on agriculture for a living" (Davis, 1953, p. 299). Historically, hobby farming has been associated with the construction of a moral landscape, of exercising one's humanity through the materiality of hard earned labor.

In today's digital realm, the act of voluntarily contributing work to the creation and sustenance of social network spaces, including that of online gaming communities, has been coined 'free labor' (Scholz, 2012). Often done collectively and driven by the democratic ideal, these laborious acts together fuel what is termed the 'gift economy.' In the heady days of Web 2.0, this was celebrated as a communitarian vision where disparate individuals came together to donate their time and energy into the making of a vibrant online space of social value. The reward is dominantly affective and moralistic, being part of a larger whole and productively contributing to a digital and cultural commons, for the people and by the people. In the book, *Wikinomics: How Mass Collaboration Changes Everything* (2006), Tapscott and Williams express deep optimism for this new labor landscape that created the much loved platform of Wikipedia. This digital domain continues to serve as an example of mass free labor manifesting in a new world vision and altruistic humanity. It serves as proof that one can organize without an overarching organization dictating work terms and conditions. Another book that fed into this frenzy of the time was Charles Leadbeater's *We-Think. Why Mass Creativity Is the Next Big Thing* (2007). Through this lens, consumers can also be producers, and leisure morphs into forms of work. Hence, from this perspective, hobbyists, part-timers, and dabblers are at last legitimized in their efforts as new media topographies provide them ample opportunities to exercise their amateur expertise and talent for the market (Howe, 2006).

Problems arise when such unremunerated and ideal efforts unwittingly become the resource and profit for companies. For instance, what happens when hobby farmers transform a piece of open space into a productive landscape and find companies and the state usurping their efforts? These virgin territories have been carefully nurtured to serve as an alternative vision to the corporate agricultural model of mass commercialism and disengagement with the land. Similarly, in the Internet context, we are noticing the monetization of free labor, a type of work that serves the corporate bottom line. The fruits of such collective cultural labor "has been not simply appropriated, but voluntarily channeled and controversially structured within capitalist business practices" (Terranova, 2000, p. 39). Time and again, we see the masses waking up to the fact that their labor has in fact generated profit for someone else. Understandably, this creates a backlash. As early as 1999, American Online (AOL) thrived on the dedication of 15,000 'volunteers,' who painstakingly contributed to the design and management of their digital platform. Over time, a feeling of being in a 'digital sweatshop' permeated this group culture, resulting in a request to the Department of Labor to investigate whether AOL owed them back wages for years of free chat hosting. This is not the only example of a platform that gained value through unpaid inhabitants toiling to shape their cultural space. Recently, we see major tensions on several digital leisure platforms. From Couchsurfing to Second Life to Flickr, the collective that once enthusiastically nurtured these digital landscapes are emerging as a labor union of sorts, demanding that their efforts and membership be recognized. The long-standing community of Couchsurfing, a global network of people who host each other when traveling and organize various social events, recently had tensions flare as some of their key members were censored for being too critical of the new design. This resulted in the Facebook page 'Censorship on Couchsurfing.' While, indeed, most of its six million members are unaware and disengaged from this protest, it is well understood that it is this core membership that sustains its grassroots spirit. In 2011, this nonprofit organization became a venture-capital-backed start-up company. From the backpacker reciprocal culture, it now feels more commercial oriented, much like Facebook and others that push their corporate agenda to increase profit.

This commodification of communalism is just one aspect of the story. Say, for instance, hobby farmers do decide to allow pleasure and profit to live side by side. There is still the issue of what their rights and responsibilities are as part-time workers in the larger agricultural politics. This is in line with questions regarding the recent trend of crowdsourcing through platforms such as Mechanical Turk. Within this web-based work environment, people with hobbies and skills come to fulfill micro tasks that are offered by a plethora of companies. While they are compensated, they are underpaid and are not protected by the usual institutional structures of mandatory health insurance and minimum wage (Scholz, 2012).

To these currently pressing issues, Jose van Dijck and David Nieborg (2009) lend a nuanced perspective that is neither indiscriminate condemnation nor blind celebration. They critically analyze the way businesses have leveraged on these social impulses and the desire to contribute to a larger whole. In doing so, they push us to ask astute questions, such as: What are cultural goods? Who owns mass produced/created cultural goods? At what levels does collaboration occur? Is innovation the sustenance of today's business? To what degree are amateurs creating value for the company? They point out that we cannot assume that businesses and consumers are in alignment in their interests and benefits. Instead, it is a constant negotiation of the shared space and cultural practice, as both parties in some sense need each other. The more indispensable a terrain becomes for the user, the more likely they will continue to (often reluctantly) pay rent via their free labor. But, because of the politics of agency, these dynamics have a way of shifting. Activism and utopianism foster new digital and urban commons that are more aligned with improved visions of social life.

Factory Gardens, Social Visionaries, and Emotional Labor

Work for work's sake is hardly inspirational. We are constantly seeking and extracting meaning from our place of employment. When we toil, we also dream. We dream of belonging to a larger cause and embedding ourselves in terrains of self-expression. The business sector has often risen to the occasion to shape this progressive social vision. In the industrial era of the 19th century, leisure was already being viewed as a potential tool to motivate and mobilize. Chris Rojek (2009), the prolific scholar on leisure studies, reminds us that the modern question of leisure in our work life was not just driven by workers demanding more freedom from their chores or statesmen with a new utopic dream to sell. It was also driven by industries that were beginning to believe that productivity was intrinsically tied to leisure practice. With institutional involvement came the codification of acceptable and unacceptable leisure and the regulation of free time and behavior. In an age of increasing urbanization, 19th-century industrialists and the state were concerned about losing control over the socialization of the working class. Providing 'normal' leisure spaces became fundamental to channeling angst and enhancing emotional intelligence, a quality tied closely to competence.

> What we now call emotional intelligence and emotional labor gradually becomes so critical and insistent on the performance of competent behavior that it compromises the notions of 'time off, choice, and freedom.' Even the practice of 'chilling' today is a performance activity that requires an acceptable setting, codes of representation and the social paraphernalia of coding and representation that signifies the suspension of pecuniary enterprise and industry. (Rojek, 2009, p. 85)

Back in the 1880s, a new type of designed green space appeared in the industrial landscapes of Europe and the United States—the factory pleasure garden and recreation park (Chance, 2012, p. 1602). Across nations, the sweeping reforms of the public domain caught the imagination of dominant industrialists, particularly that of the urban parks. They sought to hire the very architects who were instrumental in designing parks to extend such aesthetics to work arenas. The park designs of two companies—Cadbury Brothers at Bournville, U.K., and the National Cash Register Company, Dayton, Ohio—became role models for the development of corporate recreation spaces that continued to be influential through the 1960s. Viewed as 'recreational welfare capitalism,' the efforts of carving park spaces around factories were seen by corporate visionaries to add "economic, social and cultural value to the company by contributing to a more healthy, stable and productive workforce and enhancing the company's profile in the local and public realm" (p. 1603).

This modeled the labor landscape to resemble public leisure grounds and accommodated the plural needs of the workers, including those of women and their children in the factory place. Thereby, playgrounds were built near the factory. Small promenades were created for workers to rest and take a stroll and green niches allowed for picnicking and socializing. Much like how corporate blogs and Facebook pages signal a new digital corporate culture, factory gardens signaled the rise of modernity in the workplace. These intervening moments of rest paid off. Factory parks gained evidence of economic gains attributable to increased worker productivity, and leisure within these spaces began to be viewed as more of a necessity. Likewise, employees were beginning to see it as their fundamental right. Some companies went as far as to involve the factory workers in the planning and design of these park structures. This co-creation was meant to symbolize the progressive nature of the company and create a sense of worker ownership, loyalty, and personalization that was believed to benefit the bottom line.

Leisure space can enhance productivity in these domains. It can also instill personalization, immersion, and company loyalty. We witness this vision in the embracing of social network sites and blogs in today's digital corporate territory. There is belief that social media spaces open up new ways to foster social capital, collaboration, and bonding among employees in the workplace (Gely & Bierman, 2007). This comes at a time where there is deep concern about rampant social isolation felt at work. Because of a marked decline of peer support, social isolation is known to have increased substantively since the 1980s: "in 1985, about thirty-percent of people had at least one confidant among their co-workers. That proportion fell to eighteen-percent in 2004" (p. 297). In fact, as more time is being spent at the workplace, virtual networks have come to be seen as extended corporate spaces to foster employees' social connectedness and sense of community. The goal here is to nurture emotional well-being, which is seen as fundamental to the larger functioning of the employee. Take, for instance, the trend of social isolation in the United States:

> As Americans are marrying later, divorcing more often, and living alone more, work may be becoming the new center of American community, and we may be transferring our community ties from the front porch to the water cooler . . . There is hope that internet technologies—and blogs in particular—can decrease social isolation in today's workplace by strengthening weak ties between co-workers. (p. 299)

Expectations for these digital platforms are growing: employees are expected to reach out and connect with one another, facilitate a collaborative corporate culture, make business processes more efficient through outsourcing and recruiting, and improve employee training and general communication across the workplace (Kaupins & Park, 2011; Leader-Chivee & Cowan,

2008). These platforms are seen to offer opportunities for employees to demonstrate their intellectual capital and become visible to the management. In turn, corporate leaders can gauge employee motivation and satisfaction faster and respond more quickly to labor unrest. Twitter, for instance, has gained a soft spot with many corporate executives as it serves as a friendlier platform to monitor and track projects and share critical and timely knowledge and expertise among employees (Rapoza, 2009). Furthermore, there is hope that sustained engagements within these platforms can strengthen the organization's culture and deepen loyalty to the company brand. Apple is an excellent case in point: its blog and Facebook group, called Apple Students, have demonstrated tremendous success in maintaining a fan base, pushing the boundaries of what is possible in the branding of a company.

However, we should not forget that, initially, companies did not enthusiastically embrace digital leisure platforms, particularly at their onset. Even in the industrial era, there were strong social movements and grassroots activism by workers fighting for their freedom to leisure within the labor domain. The hired hands gave great weight to small measures, such as taking breaks during work to relax and socially connect with others. Work life was a substantive part of social life and was thereby seen as necessary to expand the humane boundaries by blurring labor and leisure needs. This created substantial pressure for social reform. Some companies were ahead of the curve, recognizing that this was not necessarily a threat to production. Rather, if it were strategically constructed, it could be harnessed for their economic ends. The win-win solution was their new mantra. Coming into alignment with the workers perspective and catering to their emotional need for leisure was seen as a smart move. Today, 65 percent of employees in Europe report that their everyday work life includes social networking. In general, two-thirds of employees in Europe feel that their companies are more transparent and more open because of the adoption of social networks. By country, Germany has the highest adoption rate and Great Britain the lowest (McGrath, 2010). In the United States, more large companies than small and medium enterprises (SMEs) use social media tools.

While the business sector at large recognizes the opportunities to use new media platforms, there continues to be deep concerns. These leisure grounds, be it within the factory gardens or digital networks, can also invite unwanted behavior, such as the formation of labor unions and company sabotage by the sharing of trade secrets, intentionally and unintentionally. Thereby, there is ongoing surveillance of employee activity and behavior within these supposed free social spaces. For instance, most companies in the Fortune 500 are taking advantage of the opportunities of Facebook. However, if we delve deeper, this is taking place at a peripheral level as corporations are posting mainly news releases and mission statements, being very careful of the nature of information being shared within these spaces (McCorkindale, 2010). More than three-quarters of these Facebook pages did not have any recent news or updates in the mini-feed. A recent study found that only 37 of the Fortune 500 companies maintain corporate blogs; most embrace a conventional one-way communication strategy (Cho & Huh, 2010). Employees here seem to police themselves, either by barely participating or by superficially engaging within these domains. In 2008 and 2009, phishing attacks on social networking sites soared to 164 percent. In a survey with senior marketing executives, almost 20 percent of them reported being victims of online scams and phishing attacks aimed at hijacking their company's brand names. Hence, while leisure networks are seen as important in the shaping of contemporary business cultures as decentralized and emancipative spaces for labor, there are undoubtedly challenges (Hardt & Negri, 2001).

The construction of these corporate leisure domains is hardly an exact science. The manipulation of workspace has utilitarian and symbolic repercussions: hierarchies and linear thinking can be embedded in the design and shaping of business spaces and can lead to a culture of control and privilege efficiency over creativity (Daskalaki, Starab & Imasa, 2008). Earlier research on organization culture has focused on how spatial arrangements and physical architectures can reveal aspects such as power relations, company values, and management styles (Henley, 1977). Research highlights the emotive aspects evoked by corporate structures as essential to their design (Urry, 1995). The aesthetics of corporate terrain can serve as powerful emblems of the private sector and the state (Guillen, 1997). For example, in the 1970s, modernist architecture signaled a new era of scientific advancement and elicited faith in the new corporate and state agenda. In later studies, however, there has been more emphasis on the co-creative approach to these spaces, giving more weight to employee and customer agency:

> These behavioral or functionalist approaches have been challenged by more constructivist views that utilize the notion of 'appropriation' to demonstrate how users of space participate in giving meaning to a space. That is, according to constructivist approaches, the individuals do not only use (or populate) space but also co-construct it and in effect have opportunities to subvert or divert it from its pre-conceived basis. (Daskalaki, Starab & Imasa, 2008, p. 50)

Of course, these relations shift and evolve as social engagements, policies, and economies transform, compelling us to view these spatial enactments as dynamic. Hence, the focus should be less about spatial outcomes on corporate culture and more about recognizing the nature of boundaries formed between the leisure and labor realm, employee and employer interaction, and the spectrum of policy interventions; all of these come together to create an existing organizational culture. While spatial structures of companies can be prescriptive, homogenous, dominating, and regimented, users of such space have the potential and ability to defy, play, circumvent, and modify such terrains to create a new space different from the intended corporate design (Legge, 2005).

To conclude, new media technologies offer new ways of spatializing living, working, and playing. Historically and until today, leisure is associated with "constructs such as freedom, release, fun and choice; work with constructs such as compulsion, routine and restriction" (Guerrier & Adib, 2003, p. 1399). As mentioned, these strong demarcations between work and play have been defined as a product of the industrial age. It is not new to relate work productivity to the physical environment. From blue-collar environments such as factories to the white-collar desk job, it is now commonly understood that the way workspace is organized has a social impact on employee performance, attitude, and teamwork. Granted, these new media topologies can improve communication and cooperation between workers, but they also introduce new ways to control and divide labor; "the capitalist mode of production is characterized in a fundamental way by the contradiction between competition and cooperation" (Hermann, 2006, p. 65). Nowadays, more organization's networks are becoming highly moderated and monitored within what constitutes as corporate 'walled gardens,' protective enclaves for corporate activity (see Chapter 4). Generally seen, large companies focus on internal social networks while SMEs use more external social networking tools. Hence, rather than oppose this growing and popular trend, many companies are now examining a range of ways in which they can architect and regulate leisure networks at the workplace that can satisfy both the employee and the employer.

CORPORATE INCUBATORS AND NETWORKING HUBS

Technopoles, Converging Ecosystems, and Open Innovation

The organization of space is not just within a specific company but also among different corporations. For decades now, companies from a similar sector cluster together within an urban commons, often as part of a state agenda. This public and private cooperation is based on the belief that connectivity facilitates the circulation of knowledge, seen as fundamental in promoting regional and economic development. Castells and Hall's classic book, *Technopoles of the World: The Making of 21st Century Industrial Complexes* (1994), identified this significant trend early on, highlighting the growing similarity in design of these workplaces, embedding the walled garden tradition:

> To build a community of researchers and scholars, isolated from the rest of society—or at least from its vibrant urban centers—is an old, well entrenched idea; in western societies it goes back to the medieval tradition of monasteries as islands of culture and civilization in the midst of an ocean of barbarians. The construction of such privileged,

secluded spaces is intended to signify a certain distance from the day-to-day conflicts and petty interests of society at large, potentially enabling scientists and scholars to pursue their endeavors, both detached from—and independent of—mundane material concerns. Simultaneously, the internal closure of the space is supposed to spur the cohesion of intellectual networks that will support the emergence, consolidation, and reproduction of a scientific milieu, with its own set of values and mechanisms to promote the collective advancement of scientific inquiry. (p. 39)

Known as research, science, industry, technology, and/or information parks, they became a common presence, particularly in the 1980s. In the recent decade, Asian countries such as Taiwan, South Korea, Hong Kong, and Malaysia have imported this park model wholeheartedly and have been successful in attracting foreign investment and promoting growth of knowledge-based industries in knowledge-based countries. Take, for example, the International Tech Park (ITP) in Bengaluru that opened in 2000. ITP is a joint venture involving Tata Industries, the Singapore Consortium, and the Karnataka Industrial Areas Development Board (Vaidyanathan, 2008). Located in Whitefield, 18 kilometers from the city center, it is spread over 65 acres with a capacity of two million square feet. Their incubation park facilities, popularly known as 'plug-and-play,' allow for employees to engage and interact in a seamless and creative way. These parks are also marked by their ample green landscapes spread over 300 acres or more. Cafés, gyms, yoga centers, and basketball courts dot the terrain.

These park developments are based on a vision of self-containment. This manifests in carving out open spaces that weave together recreational facilities, workstations, and conference plazas in an idyllic setting. Lawns, malls, open space courtyards, and quadrangles are some of the spatial expressions of leisure within these corporate parks. Underlying this is the belief that the balance of leisure and labor can create conducive environments for innovation. They are meant to serve as incubators for ideas and foster networks and synergies with other private-sector companies. With incubation, however, comes the challenge of controlling and securing essential business secrets and business strategies:

> The ease of interaction among park enterprises is a function of the physical layout of a park, the extent of planned events such as colloquia, seminars, and so on, and the internal policy of park tenants. Park enterprises might prefer minimal, or tightly controlled, opportunities for external interaction in order to minimize labor force raiding and the leaking of innovation to competitors. (Goldstein & Luger, 1990, p. 73)

Additionally, the role of the state in these park constructions cannot be underestimated. The state has often been instrumental in this unique union of competing companies within a sector. For instance, Russian parks

(Kihlgren, 2003) were created in 1997 because of the 'Promotion of Innovation' state policy. The program was geared towards creating a network of centers in regions with high scientific, technological, and innovative potential. Similarly, if we look at China, the first national science and technology industrial park (STIP), Beijing Zhongguancun, was approved by the Chinese State Council in 1988, followed by 27 national parks in 1992 (Zhang & Sonobe, 2011). According to the Statistics Report of the China Torch High Technology Industry Development Center, there were 43,249 high-tech firms in China in 2006, with 27,293 on-park and 15,956 off-park (Zhang & Sonobe, 2011). However, they are not on the same level playing field: "while the on-park firms are clustered in STIPs, the off-park firms are scattered. Another important difference is that on-park firms are more favorably treated by the government than off-park firms" (Zhang & Sonobe, 2011, p. 3–4). Strong policy and government incentives transform these workspaces into innovative hubs by attracting the best and the brightest. This is attributable to their enticing work environment and unique privileges afforded by the state, giving them an advantage over other non-park corporate spaces.

However, in many cases, state intervention may not be sufficient. The mere clustering of companies does not automatically lead to globally competitive products and services but they do serve as hubs for general business support. Kihlgren notes that "science parks should be seen more as service organizations providing a range of business support facilities to technology-based firms than as centers of scientific excellence" (2003, p. 75). In fact, this causal link of architecting corporate spaces as 'park' spaces to foster innovation has been viewed as problematic for quite some time now. Massey, Quintas, and Wield (1992) critiqued this trend early on as 'high-tech fantasies,' accusing the state and corporate entities of being delusional in their beliefs that mere proximity could lead to innovation in these industrial parks. When examined closely, even the celebrated park template of the Silicon Valley in California and the Japanese technology parks demonstrated that besides proximity, state policy, community culture, urban trends, and other context specific factors contribute to whether corporate parks succeed or fail:

> Today Silicon Valley owes its prestige and success not so much to the 'entrepreneurial spirit from below' as to the presence of information technology (IT) giants like Intel, Cisco, Hewlett Packard and Sun, well-funded research institutes, and a network of assemblers with an extraordinarily high percentage of migrants and women workers employed at low wages under poor working conditions. If there is an overarching characteristic of Silicon Valley, it is the absence of unions or other forms of organized social resistance. From this point of view, Silicon Valley can indeed be seen as a role model. (Hermann, 2006, p. 67)

While several attempts to replicate this model have failed across the globe, some noteworthy successes are the Chilecon Valley in Santiago, Silicon Wadi in Tel Aviv, and San Pedro Valley in Belo Horizonte in Brazil. What these clustered corporate topographies have in common is their overt effort to foster a work culture that is infused with social dynamism. For example, they have regular 'hack days' in San Pedro Valley and there are social outings for start-ups with established tech gurus. In the case of Japanese technology centers such as Izumi Park Town, Tsukuba, and the Kansai Science City, their success is attributed to the tapping of the innate leisure needs of employees (Forsyth & Crewe, 2010). There is a yearning for natural surroundings and a connection to the land. Studies found that there are three types of workers: *the pragmatists*, who demand convenience and urban-centric workspaces; *the naturalists*, who prefer open and natural settings; and *the community-focused people*, who value the idea of nurturing networks and relationships. Contemporary corporate spatial design today particularly targets mobile, highly educated globetrotters who, because of their constant movement, deeply value predictability and access to international amenities. Green space architectures thereby are a balance of manicured as well as naturalistic areas. Planners and designers explain their approach as they translate this vision:

> create an open atmosphere promoting innovation and adventure and tolerating failure for talented, well-trained, and highly mobile staff' . . . to 'enhance the visual aesthetic value of the software park by taking ecological landscape planning into consideration' in terms of protecting natural features and vegetation, buffering the site visually, turning non-conforming uses into green spaces, and controlling the microclimate with trees. (p. 180)

When shifting our attention to the digital commons, one might ask, how does this cohabitation of different companies translate into shared virtual networks? Is there a clustering effect going on within today's information infrastructures? What are the concerns related to the converging of knowledge via online platforms? And what indeed is the role of the state in fostering open innovation systems online? There is still tremendous experimentation going on with business models in the digital domain; businesses by and large cannot afford to not cooperate. Corporations are just beginning to grasp and capitalize on the enormous potential of converging new communication infrastructures in this growing leisure-oriented economy. They now recognize their indelible dependency on each other as consumers demand flexible infrastructures, exercising their right to move from one platform or device to another.

Whether it is access to gaming apps or watching our favorite television show on different competing devices, or the ability to take our friends with us to the new and promised lands of social network sites, competitors are compelled to cooperate. In the recent decade, there is a growing focus on such interconnectivity or what is termed as 'digital ecosystems.' This term is inspired by natural ecosystems that describe "a set of distributed, adaptive, and open socio-technical systems. Being parts of such ecosystems, individual persons, public and private organizations are becoming increasingly dependent on each other" (Krogstie, 2012, p. 137). Companies today recognize that it is not just information exchanges that are the building blocks of this economy but new knowledge creation enabled by collaborative digital networks. After all, an increasing fraction of value creation in this contemporary economy comes from knowledge work.

The monopolies of today, particularly the media and telecommunication industry companies, such as Apple, Google, and Microsoft, can no longer isolate themselves within their own structures (more about media ecosystems in Chapter 6). Partly this is a market necessity and partly it is a sociopolitical imperative. With the acceleration of interactive technologies in the last decade there has been a push for media convergence. This has become quite a buzzword, backed by the frenzied proposition of merged content and the blurring of producers and consumers and private and public interests. Production, delivery, and processing are seen to be undergoing a radical change. And, indeed, in the 1990s, companies such as Time Warner, Sony, and News Corp demonstrated that to survive, it was necessary to foster complementary assets to create a digital commons among themselves (Carpenter & Sanders, 2007).

That said, as with corporate parks, not all synergies result in market capitalization and innovation (Ali, 2013). There are a range of factors that contribute to the success or failure in building these digital corporate landscapes. Cooperative initiatives of merging web structures are risky and sometimes directly pose a threat to the organization's competitive advantage. A delicate balance is required that keeps in mind the long-term versus short-term benefit and the tactical advantages in exchange for short-term loss in competitiveness. In fact, these digital ecosystems can be deeply confronting as they bring cooperation head-to-head with direct competition, making this an environment for corporate 'frenemies' (Gupta, Kim & Levine, 2013). For instance, within such cooperative arrangements, Samsung and Google partnered together to stem Apple's dominance in smartphones. Simultaneously, Google acquired Motorola and developed sophisticated handsets that compete directly with Samsung. They did this by approaching other partners in the industry to ensure their Android devices would check Samsung's rise in the market. However, as Ali (2013) remarks, in spite of these obvious conflicts:

> partners understand that the cooperative arrangements will indeed fuel competition among them but that they are needed to cope with market challenges . . . Lastly, the strengths and weakness of partners are in fact complementary. Apple and Samsung, for example, have different

strategies. Apple dominates the market in the industrial world, while Samsung is relatively more popular in the developing economies. These differences, along with formidable strengths, have rendered quiet cooperation a better strategy than all-out war for some time to come. (Ali, 2013, p. 1–2)

As for the role of the state, slowly but surely we witness some governments making public their massive citizen databases. There is a belief that shared knowledge can lead to pioneering ways of organizing the public and the private sector. According to the recent Digital Agenda for Europe, "Smart use of technology and exploitation of information will help us to address the challenges facing society" (European Commission, 2010, p. 27). In 2005, the government of India passed the Right to Information Act (RTI), institutionalizing the citizen's right to access government records. According to a recent article in The Economist, the Obama administration has declared that going forth, all data created or collected by America's federal government must be made freely available to the public. Such vision stems from the belief that information from diverse and seemingly unrelated sectors can converge to produce new ideas and ways of practice in both the private and the public sector.

Pollution numbers will affect property prices. Restaurant reviews will mention official sanitation ratings. Data from tollbooths could be used to determine prices for nearby billboards. Combining data from multiple sources will yield fresh insights. For example, correlating school data with transport information and tax returns may show that academic performance depends less on income than the amount of time parents spend with their brats. ('Open Data: A New Goldmine,' The Economist, 2013)

Hence, depending on the state's outlook and impetus on open innovation, we can gauge the effectiveness of digital ecosystems.

Overall in these last few decades, corporate parks have exponentially sprung up across nations and are a popular option of organizing workspaces. This is because of the persistent belief in fostering innovation and business networks. Similarly, digital ecosystems are becoming the norm in this new media age. The embedding of leisure infrastructures is seen as necessary to transform zones of work into zones of inspiration, blending work with play. The creation of corporate clusters as cooperative spaces has come to the forefront with an emphasis on promoting entrepreneurship and innovation. The state is not far behind in orchestrating these seemingly contradictory networks; there is a long-term vision of how such merged infrastructures, both materially and virtually, can transform the information economy. However, mere linkages and connectivity are not sufficient to foster innovation and mutual benefit. A complex set of factors, such as political momentum, long-term visionary thinking, a perceptive shift on leisure and labor cultures, and newly erected legal boundaries on consumer rights and activism, contribute to the outcomes of these corporate parks.

When addressing the fierce competition in this globalizing economy, corporate arenas do acknowledge leisure landscapes. Part of the appeal for play within work is attributable to the current belief that leisure terrains foster a more inclusive organizational culture, and shared recreation enables cooperation and creativity. However, the materializing of such spaces remains a challenge, because of the evolution of new technologies, legal structures, and consumer behavior. Certain corporations do not partake in these new business flows. Partly this is attributable to their innate work culture: They are driven by narrow visions of profit and are in general risk-averse, especially without sufficient evidence linking leisure directly to productivity. At times, even when corporations do engage, they do so at a peripheral level. The habit of replicating the known and established industry park models, thereby resulting in the 'Disneyfication' or cookie-cutter treatment of workspaces, defeats the very intent for creativity through spatial design.

We see this attitude can uncritically transfer to digital ecosystems. New platforms appear to resemble the old online and are thereby doomed to failure. Sometimes, this could be because of the over-involvement of the state, policy makers, and planners in the creative process. Also, we need to go beyond the Western model as we cannot ignore the range of emerging markets embracing and appropriating these models. The global South is yielding diverse outcomes because of their unique specificities such as policies, infrastructures, demographics, and market mechanisms. We also need to re-examine the causality of corporate leisure spaces leading to innovation, entrepreneurship, strategic networks, and regional development.

Work Cafés, Sociality, and Entrepreneurship

The problem with corporate parks is that they are corporate. The nature of the beast is such that these domains lean toward standardization, uniformity, and predictability: anathema for creativity. Even in the urban center, where they extend their reach into the public sphere, the result is often the conforming of public leisure space.

> Downtown rebuilding efforts not only change a city's skyline but also transform its urban form. A distinctive feature of the new downtown is the variety of open spaces created through private enterprise: plazas, paseos, gallerias, roof gardens, and arcades . . . This can be attributed to the fact that the goals of commercial and corporate developers are similar everywhere and these are the goals expressed and served through design. Indeed it is clear that increasingly a great deal of attention is given to developing a certain mood for the space, to promoting a theme-park-type setting, to packaging and advertising the product,

and finally to managing and maintaining the theme park environment. The planning and development of a modern downtown office complex is not unlike what is involved (on a grander scale) in the planning and design of Disney World or Universal Studio. (Loukaitou-Sideris & Banerjee, 1993, p. 11)

Over the last few decades, these downtown plazas have been examined and, given the risk-averse nature of corporations, are consistently found to be sanitized spaces. There is a lack of creativity within such spaces, having a potentially adverse effect on entrepreneurship and innovation. There is a curtailing of architectural freedom in creating spaces that cater to the ambiguities and improvisational aspects of human action. This freezing of corporate space into predictive styles is not surprising given the tendency of corporations to imitate success through replication. These spaces are furnished for specific demographics, mainly white-collar workers, negating a diversity that can lend inspiration to corporate activity.

In the last decade particularly, the café has become the terrain of choice for a number of entrepreneurs to share ideas and networks. This fits with a larger vision of the state to promote the making of innovative cities to attract entrepreneurial and creative types to inhabit these spaces and transform the culture of the cityscape. Again, this perspective has historically played out in a number of creative cities of the past. In Vienna, around the 1900s, the city attributed its vibrancy to the 'café factor' (Hospers, 2003). Countless Kaffeehäuser were open early morning to late night and served as a meeting place for local creative minds. This was viewed as an inspiring environment, where "a number of 'new combinations' emerged while drinking a cup of Wiener mélange, or the local beer" (Hospers, 2003, p. 265). The relation between work and the café culture has been recently explored quite extensively, particularly through the emergence of the mobile worker who views the café as a flexible area for productivity. While away from the office, employees are able to carve out a temporal workspace. This is not to say that employees are free of work constraints when relocated to a leisure domain. Even when away from the organization's prying eyes, employees often self-regulate and discipline themselves within these settings, underlining the influence and reach of corporate culture into public leisure domains (Laurier, 2008).

It is understandable to see the mirroring of this phenomenon in social network sites such as LinkedIn and Facebook. These sites are designed for connecting and building creative capital within a more leisure-oriented setting. A study by Meredith Skeels and Jonathan Grudin (2009) reveals certain motivations for professionals to embrace these sites. There is a common perception that these are informal and disarming domains to maintain weak ties with other professionals. Through sharing of pictures, hobbies, travel, and interests, people report that these spaces allow a more comprehensive outlook. They provide more genuine rapport and interactivity that is not just instrumental in nature. When it comes to entrepreneurship, these digital leisure networks are seen as a lucrative gateway in reaching out to new clients and experimenting with new ideas and markets.

The creative industry in particular—think emerging artists and musicians—has managed to leverage these platforms in a substantive way. Numerous sites such as MySpace, Last.fm, and The Hype Machine have served as creative grounds for musicians to build an online community and launch themselves (Carter, 2009). Research has gone into how online social networks influence the diffusion of music and circumvent big music industries through social contagion and the viral embracing of new products (Leyshon, 2001; Watson, 2012). Of course, this new digital sharing of music comes with a reorganization of the music industrial landscape, posing new challenges and opportunities for not just the entrepreneur but also corporations willing to embrace change. This is evidenced, for example, in the success of the iTunes business model.

In his book, *Socialnomics: How Social Media Transforms the Way We Live and Do Business* (2012), Erik Qualman is full of examples of how entrepreneurs use these platforms to test out ideas and birth new business practices. Take, for instance, Bacon Salt, a brainwave of two Seattle friends, Justin Esch and Dave Lefkow. They wondered if there could be a powder that "made everything taste like bacon" (p. 29). So they started a MySpace profile where they sought out people who had mentioned bacon in their profiles to see if there was interest in pursuing the idea. They were astonished that this small effort created a surge of interest and orders for a product that was yet to be invented.

> Word of Mouth took over from there, and as Lefkow describes it, "It was one person telling another person, telling another person. It was amazing and scary at the same time. We weren't prepared for the onslaught." The viral aspect of this experience branched into non-social media channels, and they even received a free endorsement from the Gotham Girls Roller Derby team. It's one thing to get buzz about your product, it's another thing to sell it—and sell they did. The spice that made everything taste like bacon incredibly sold 600,000 bottles in 18 months. (Qualman, 2012, p. 29)

Since then, there have been numerous examples of entrepreneurs innovating by first testing their ideas online and gauging reactions and support for their potential business plans. Organizations such as Kickstarter, the world's largest funding site that connects entrepreneurs with funders for their creative projects, have emerged to capitalize on these digital network trends. Kiva.org is another such site but for poor people in the global South trying to empower themselves and establish global networks through entrepreneurship. In essence, informal leisure spaces provide an innocuous ground upon which entrepreneurs can make new connections, share perspectives, sustain weak networks, and potentially launch new ideas, reminiscent of the material and historical culture of the café.

To conclude, there is an underlying rationale to structure corporate spaces. This manifests materially, through business parks, and virtually, through social media platforms. There is a recognizable shift in the organizational culture where the role of leisure has become more central in facilitating a employee and customer-driven approach. From initially rejecting such phenomena, businesses are now embracing and capitalizing on these blurred boundaries of work and play. They do so to maximize productivity and creativity in the workspace, online and offline. The belief that these open, shared, and relatively unregulated spaces can foster creative, out-of-the-box thinking has infused the architecting of such workspaces. After all, innovation is a way for businesses to remain competitive in this global and digital era.

However, we see evidence emerging from industry, technology, information and science parks, and social network sites that the mere proximity of companies within a concentrated space does not necessitate novel knowledge production. In fact, there can be surprisingly low levels of knowledge circulation within these so-called incubators of ideas. Partly this has to do with the need to protect against the leaking of company ideas. The struggle to balance cooperation and competition is very much a part of how the market functions. Granting freedom within protective and monitored enclaves is becoming more the norm, compelling us to be more critical of these so-called open corporate spaces. This holds true for many examples, such as Facebook pages of corporations or cafés within 'techparks.'

Another expectation that prevails is the hope that the clustering of businesses will foster strategic networks. We see this in the vigor and exponential growth of corporate blogs, online communities and Twitter accounts, to the proliferation of information parks around the world. Yet the formation of networks does not materialize unless other factors collate to influence this clustering. Such factors include policy initiatives, special facility privileges, government intervention, market conditions, and the like. Silicon Valley is a good case in point; so are initiatives online such as Microsoft's embracing of new media platforms for internal idea sharing.

The success of clustering in emerging markets has less to do with proximity and more to do with the concentration of facilities. This occurs in the midst of conditions that are generally starved of good infrastructures, giving them a disproportionate advantage. These spaces serve as symbolic environments of expertise that allow a nation to define itself within the domestic and global market. Partly it is to attract talent and create a safe and communal environment in an age of dislocation and social isolation and fragmentation.

Lastly, we need to pay close attention to how these walled enclaves relate to their outside environments. We also need to examine larger regional contexts and the ways in which they negotiate and make transparent what constitutes as corporate and public space. There is much concern that genuine public space is being lost as corporations commercialize and 'Disneyfy' common public grounds (more on this in Chapter 6). This can be seen with the rise of advertisements and data sales within social network sites and with downtown plazas becoming corporatized public grounds. Overall, we need to give credence to the fact that online and offline corporate spaces are transforming and responding to a new organization culture. This culture is driven by new expectations, economies, technologies, and global events. We also need to situate digital appropriation of leisure spaces like blogs and social network sites as extensions of historical efforts to create clustered network spaces to enhance productivity. Leisure space has come to the foreground for the shaping of corporate space. We need to approach this phenomenon in a more integrated and holistic way, giving weight to a range of social and business enactments, both online and offline.

6 Fantasy Parks
Consumption of Virtual Worlds of Amusement

I don't want the public to see the real world they live in while they're in the park . . . I want them to feel they are in another world.

Walt Disney, *Official Disney website*

The City Beautiful movement . . . [its] fascination with sumptuousness, visible order, and parks—with the monumental 'public' aspect of the city—anticipates the physical formula of the theme park, the abstraction of good public behavior from the total life of the city.

Michael Sorkin, *Variations on a Theme Park*

Because virtual worlds appear so novel and in such a constant state of change and expansion, understanding their history can be difficult. However, virtual worlds did not spring like Athena from the forehead of Zeus, full blown from the mind of William Gibson . . . [They have encoded within them] a complex history of technological innovations, conceptual developments and metaphorical linkages.

Tom Boellstorff, *Coming of Age in Second Life*

Fantasy is a very human response to real life. We seek alternatives and contrast to the typical environments we inhabit. We immerse ourselves in the implausible and never cease attempting to make the impossible possible. Fantasy transforms the mundane, everyday ritual of living space into a playful terrain that allows for new forms of social interactivity and emotive fulfillment. One of the most powerful spatial manifestations of this raw need is Disneyland, an ingenious effort at organizing fantasy into marketed reality. This empire of escapism has spread globally and now serves as an icon of a utopic leisure landscape. Within the United States, there are now more than 400 'Disneyesque' amusement parks and, if we are to look at Europe, we would find 300 such parks scattered across its terrain. This is a huge moneymaking enterprise, having generated an estimated 4.3 billion euros (5.3 billion dollars) in total revenue in Europe alone and contributing approximately 8.6 billion euros (10.6 billion dollars) to the European economy in 2008 (International Association of Amusement Parks and Attractions, 2013). Even the emerging markets have jumped on the bandwagon, in spite of their economic slowdown and continuing issues with infrastructure. At least eight theme parks have opened or are scheduled to open in West Africa alone since 2000 (Hinshaw, 2011). As their youth populations grow and demand novel terrains to experience leisure, Malaysia, China, India, and others are well down the line to embracing this new fantasy environment.

This place making, it is popularly believed, began when Walt Disney made a trip to the Chicago Railroading Fair in 1938. Dressed in engineering overalls and sitting behind a historical locomotive, he got inspired to create an environment where childhood fantasies could come alive and thrive (Sorkin, 1992). Early publicity pamphlets captured the essence of his vision:

> Disneyland will be based upon and dedicated to the ideals, the dreams, and the hard facts that have created America. And it will be uniquely equipped to dramatize these dreams and facts and send them forth as a source of courage and inspiration to all the world. Disneyland will be something of a fair, an exhibition, a play-ground, a community center, a museum of living facts, and a showplace of beauty and magic. It will be filled with the accomplishments, the joys, the hopes of the world we live in. And it will remind us and show us how to make those wonders part of our lives. (p. 206)

Of course all utopias are grounded in real-world challenges of how to finance and market such domains. This is particularly true when such propositions are astoundingly ambitious and appear unprecedented in nature. In this case, the birthing of this fantasy park came with Walt Disney offering his famous Mickey Mouse icon to the television channel ABC, fostering simultaneously the physical Disneyland and the Mickey Mouse TV series. This laid the foundation of a unique symbiotic relationship between the media industry and amusement parks that continues to pervade and thrive today.

Take, for instance, the recent Finnish digital game sensation Angry Birds: It became a runaway hit in the mobile app world with 1.7 billion downloads across all platforms (Gaudiosi, 2012). It relies on a simple plot, where players use a slingshot to launch birds at pigs stationed on or within various structures, with the intent of destroying all the pigs on the playing field. Given its tremendous success and global themed variations, the Japanese Cheery Blossom, the Go Green and Get Lucky St. Patrick's themed levels, it was not a surprise that it propelled investors to materialize this virtual fantasy into a theme park. In 2011, the Angry Birds amusement park opened with much fanfare in China's Hunan Province, enabling visitors to literally immerse themselves by catapulting giant stuffed birds at green pig

Figure 6.1 Angry Birds land in Särkänniemi, Finland.
Source: Juho Paavisto/Wikimedia Commons/Public Domain

balloons. GameSpot senior editor Giancarlo Varanini told Fox News that, "as video game brands continue to grow in popularity, there will undoubtedly be more attempts at using their built-in audience to lure more people into parks that would otherwise not care" (Carlton, 2011, p. 1).

In fact, the shaping of physical fantasyscapes by digital mediascapes is hardly a new trend. In the last few decades, there has been a sophisticated orchestration of blockbuster movies, television series, and video games making their way to the architecting of theme parks around the world. So you may find yourself grabbing a meal with your family at the Texas Chainsaw Cannibal Cook-Off located in the 'Happy Land' area of the park. Or you may find yourself screaming on the Flight of the Hippogriff rollercoaster ride on Harry Potter's Islands of Adventure. Either way, escapism from reality does not exempt you from escaping mass media sensations. In fact, it may be the prime reason that you are lured into these amusement domains. It is tempting to view this totalizing landscape of blurred reality and fantasy, of commerce, desire, and affect as somewhat novel in this digital age. However, this chapter emphasizes the historicity of public spaces of fantasy and how they were reflective of the public values, sentiments, and social transformation of the time. By looking at the precedent of such leisure spaces as well as contemporary manifestations, we can attend to the shifts in the cultural tone of society toward the notion of fantasy.

Here, 'fantasy parks' serve as a compelling metaphor for understanding digital amusement ecologies such as video games and virtual worlds. This helps investigate the complex interplay of the citizens, corporations, and the state in the makings of immersive virtual fantasyscapes. Through such an examination, we pay attention to the extent of online and offline public participation in the design and execution of these amusement topologies. The sanitized and predictive quality of these spaces threatens a homogenizing of fantasy parks. Yet there are localizing dynamics that stem from indigenous interpretation, play, and representation of generic icons of fantasy and Western-oriented mass media narratives.

THE COMING OF AGE OF MASS CULTURE

Theme parks are carnivals frozen in space and time. Since medieval times, there have been numerous fairs, vaudeville theaters, band pavilions, festivals, and other temporal spectacles of fantasy to engage the public. Carnivals have served as a safety valve for society; this was illustrated earlier on in this book through Bakhtin's *carnevalesque*, a feast of subversion through humor and satire and an opportunity for the masses to transform the sacred to the profane. This contributed to the sustenance of the ruling class and imperial powers of the time by channeling the energies of the masses in creative and sacrilegious directions. Shortly afterward, the 17th century witnessed the rise of 'pleasure parks' throughout France and Europe (Clavé, 2007). These were some of the first permanent sites for outdoor entertainment, designed as gardens that allowed for dancing, outdoor theater, staged spectacles, and basic amusement rides. The pace of these amusement parks was slow and marked by leisurely strolls. However, by the end of the 19th century, public needs had changed to a more risqué, fast-paced, and thrill-oriented feeling. This was in the midst of a major social transformation, where mass migrations of people from rural America and Europe were leaving for industrial jobs in the city. This explosive urban phenomenon brought together a massive concentration of immigrants with their own traditions and values but with one thing in common: the desire to assimilate. This period of time experienced a profound cultural change in the social fabric. As work conditions became more stabilized, it was marked by an increasing demand to consume leisure spaces and immerse in alternative social activities. The era of mass culture had finally arrived (Leach, 1993).

Few understood this better than Frederic Thompson, the showman of Coney Island. What appeared to be a frivolous need of the public, Fred Thompson took seriously and, together with his partner, Elmer S. Dundy, started an amusement park in Brooklyn, New York, that would revolutionize public fantasy terrains forever. He recognized the deep-seated need to immerse in creative and phantasmagorical worlds, without the inconvenience of unpleasant surprises. In September 1908 in an interview with *Everybody's Magazine*, Fred Thompson expressed his vision of the manufacturing of this new mass culture of fantasy:

> This spirit of gaiety, the carnival spirit, is not spontaneous, except on extraordinary occasions, and usually its cause can be easily traced. Almost always it is manufactured [. . .] In big amusement enterprises that appeal to the masses the spirit of gaiety is manufactured just as scenery, lights, buildings, and the shows generally are manufactured. That's the business of the showman to create the spirit of gaiety, frolic, carnival; and the capacity to do this is the measure of his mastery of the craft. Nearly all the big national expositions fail financially because, while in essence they are really nothing but shows, almost never is one run by a showman. When people go to a park or an exposition and admire the buildings, the exhibits, and the lights without having laughed about half the time until their sides ached, you can be absolutely sure that the enterprise will fail. (p. 379)

This insightful perspective highlights the marriage of entertainment with commerce in the making of amusement parks. This perspective finds resonance in today's multibillion-dollar gaming industry. In *The Ultimate History of Video Games* (2010), Steven Kent describes the mechanics of some of the major organizations instrumental in fostering such fantasy platforms. However, he points out that as late as the 1970s, only a few companies truly leveraged on video gaming as a major industry. In 1972, Nolan Bushnell, an electronics engineer from Northern California, and his partner, Ted Dabney, incorporated Atari with an initial investment of 250 dollars each. Within 10 years, Atari grew into two billion dollars a year entertainment industry, making it the fastest-growing company in U.S. history. This was not dumb luck. Much like Fred Thompson and Elmer Dundy, they were quick to recognize that immersive experiences in fantasy had currency and could be tremendously lucrative. Part of their success can be attributed to the ingenious way in which they transformed games such as Ping-Pong, designed for television consumption, into a more multiplayer and interactive electronic medium. Today, the competition is fierce as new mobile apps serve as new avenues of profit, stimulating a host of designers, computer engineers, marketing gurus, and business managers to craft the right balance of spontaneity and surprise, of engagement, immersion, and challenge within gaming worlds.

These pioneers in the amusement business, whether the fantasy park of Coney Island or the gaming empire of Atari, were not just sensitive to new business opportunities. They were tuned to the significant cultural shift taking place around them. In his book *Amusing the Million: Coney Island at the Turn of the Century* (1978), John F. Kasson points out that in the realm of commercial amusements at the time, Coney Island was exceptional as it uniquely catered to "the new cultural order, helping to knit a heterogeneous audience into a cohesive whole" (p. 4). Kasson argues that unlike the 19th century, which was governed by Victorian values of moral integrity, self-control, earnestness, and industriousness, a new cultural tone was emerging. It was about accepting and even encouraging individualism, self-indulgence, and hedonistic pleasure. Where once it was believed that leisure should have a constructive value, the time had come to look at leisure as a way of pleasure for pleasure's sake. Past engagements with fantasy viewed audiences as passive consumers. However, with Coney Island, the new pleasure-seekers were no longer mere spectators but were intrinsically involved in the larger theater of fantasy. Such changing notions of audience participation are also seen in the epochal shift from Web 1.0 to Web 2.0. The narratives of Web 1.0—of entertainment dissemination to the consuming public—have shifted to the now much lauded Web 2.0 of user participation and user construction of their own fantasy content. We have come a long way from the beloved Pac-Man of the 1980s, where users followed carefully designed pathways of gaming. Now, we turn to World of Warcraft, where, for instance, users can construct entire virtual worlds of their own choosing and take on diverse avatar forms of self.

Furthermore, the stigma that stemmed from massification connoting lowbrow participation was also undergoing a metamorphosis, as new alliances between "members of the cultural elite and commercial tastemakers made the hegemony of the genteel culture possible" (Kasson, 1978, p. 16). Here, amusement parks served as laboratories of this new mass culture, fostering new acceptability in public desires, demands, and behaviors. In other words, these venues served as new cultural institutions that challenged past Victorian values of social order and conduct, inspiring Kasson to anoint these theme parks as 'harbingers of modernity.' At this stage in the book, we are well aware of the exhaustive and enthusiastic discourses around the amateur audiences of popular culture: the prosumers that participate in the architecting of their fantasy worlds. The crowd, once a derided entity, is now celebrated for its wisdom in this Web 2.0 age. In fact, the lifeblood of virtual worlds such as World of Warcraft is sourced from user activities, intimate participations, and their affective approach to these platforms. While indeed class distinctions continue to affect choice of gaming and sites of inhabitation and engagement, the digital leisure industry continues to stay clear of any elite label: their commercial success depends on the massification of fantasy products.

This rise in the status of the masses comes with larger social repercussions. Coney Island and other such novel amusement parks of the time triggered challenges in social norms, particularly regarding gender relations. These times were marked with strong gender divisions of labor and leisure, of women being wives and homemakers while men were breadwinners. Women were generally confined to the privacy of their homes, while men engaged in the public domain. However, part of what made Coney Island so successful was in its breaking away from not just the routine of daily life but also from social roles and customs. This amusement park allowed subversions from the typical role-play of gender, and women particularly harnessed this opportunity. They were able to shelf their daily chores and domicile and instead become more laid-back and indulgent in their pleasures and desires. Also, contrary to the social protocol of keeping a decent distance between men and women, here they could get on rides together, romantically embracing as they spun around in the Ferris wheel. This new fantasy environment enabled loosely-tied networks among strangers and an odd permissibility of acquaintanceship and intimacy not easily possible in other social situations.

Not coincidentally, when we shift our attention to virtual worlds, role-playing is fundamental to these domains. In the classic essay by Sherry Turkle on playing in the Multi-User Dungeons (MUDS), she proclaimed early on that these virtual worlds of fantasy break social rules through the embrace of multiple selves of our own making:

> In the MUDS, the projections of self are engaged in a resolutely post-modern context. Authorship is not only displaced from a solitary voice, it is exploded. The self is not only decentered but multiplied without limit. There is an unparalleled opportunity to play with one's identity and to 'try out' new ones. MUDS are a new environment for the construction and reconstruction of self. (1994, p. 158)

Today however, even the most optimistic of scholars curb the enthusiasm for such limitless possibilities. This occurs as corporate interests have become powerful and dominant actors in the makings of these fantasy worlds, both physical and virtual. While encouraging the perception of participation in this experiential economy and umpteen choices in fantasy, contemporary leisure environments are getting more structured, more formulaic, and more walled in. We find ourselves navigating such platforms with the benign guidance of personally tailored advertisements and brands embedded in our playgrounds. One of the strongest critiques of this trend comes from Todd Gitlin, in his book, *Media Unlimited: How the Torrent of Images and Sounds Overwhelms Our Lives* (2002/2007). Gitlin embarks on capturing the cultural tone of our society in today's digital age. He argues that such self-indulgence and hedonism can foster a sense of disposable experiences, peripheral affective states, and, overall, a hollow sense of existence. This seemingly endless choice of fantasy is really a limited diversity of frustrating paradoxes: "Whatever the diversity of texts, the media largely shares a texture, even if it is maddeningly difficult to describe—real and unreal, present and absent, disposable and essential, distracting and absorbing, sensational and tedious, emotional and numbing" (p. 7). Regardless of gender, class, or race, the heterogeneous public is unified in the amount of time and their immersion in these media experiences.

However, this supposed democracy of culture is in reality, the partaking of heightened pleasures of numbing proportion. The digital commons, Gitlin underlines, have become collective spectators of an endless and mindless media torrent. They are the 'fugitive publics' of manufactured escapism, detribalized into personal zones of peripheral enjoyment. In this contemporary age, the mediascape seems to be constructed out of stereotypes, where girls can be Pocahontas and boys can be Tarzan.

THEME PARKS, BRAND EMPIRES, AND DIGITAL CULTURES

The Disneyfication of Fantasy Space

While at the time Coney Island was radical in recognizing the immersive potential of fantasy parks, Disneyland was a pioneer in the global scaling of these leisure landscapes. It is not surprising that scholars early on were quick to use the term 'Disneyfication' to capture the globalizing of fantasy and mass cultural homogenization that pervades the media ecology (Cameron & Stein, 2002). These parks are justifiably seen as empires: they are built on the foundation of decades of pervasive Hollywood narratives, allowing them to be read and related to by one and all across the globe. With mass merchandizing of all things Disney, this majestic machine has entrenched in the minds of tourists the script of fantasy that starts and ends with this franchise. And a powerful script it is. Synergy between their television shows, motion pictures, theme parks, and products enables a seamless and consistent experience of entertainment. One can relive an experience again and again by consuming different spatial manifestations of the Disney Corporation. Strategic partnerships have been and continue to be made among media giants to enhance this global media flow of fantasy. Time Warner, Viacom, Blockbuster, and MCA are examples of companies who combined their resources and holdings to boost their promotional activities on this common theme park platform (Clavé, 2007). Historical partnerships between Disney and MGM and Universal succeeded in merging their film and animation specialty to create the modern-day spectacle in multiple geographic and digital formats. Or take, for instance, MTV, which came together with Nickelodeon to capitalize on their live musical products and young celebrities to contribute to the makings of Paramount Parks.

Hence, when assessing gaming platforms, we cannot view them as independent terrains but as spatial extensions of already well-entrenched corporate media ecologies. They serve as novel and vital contexts for opportunities in commercial clustering. Fundamental to the workings of this global operation of fantasy is the licensing economy that pervades across all mediums, both online and offline. Take, for instance, the license for the film *Lion King*, one of the most lucrative media events of our time. Licensed images and merchandize are the heart of its phenomenal success

(Mitrasinovic, 2006). While the film's box office estimated its initial run to be around 267 million dollars, the main revenues came from the licensing of merchandized themes, which accounted for as much as one billion dollars. SNES, NES, Game Boy, PC, Sega Mega Drive/Genesis, Amiga, Master System, and Game Gear have licensed their images, music, characters, and scripts for the digital game. The game traces the life journey of Simba, from a carefree cub to a young lion who eventually battles with his uncle Scar for the forest title of king. In viewing this complex media flow, the notion of boundaries between theme parks and the media giants make little sense in an all-encompassing terrain.

Drawing attention to this corporatization of spaces of fantasy consumption, Mitrasinovic makes a case that this is more than just a loss of public space. In fact, we are in the midst of a significant cultural transformation that is affecting our architectural surroundings, cultural expressions, and social relations. He addresses this phenomenon as 'totalizing landscapes,' arguing that there is a distinct military logic that dictates these realms. This logic provides a sophisticated and efficient framework to operationalize the reproduction of everyday activities within these fantasy parks, virtually and materially. For instance, detailed feasibility studies, attendance projections, and (online and offline) traffic analysis run this machine. Such an approach allows for this model to be ubiquitous and transferrable to not just diverse international contexts but also old and new media platforms. Hence, this Disney-style public space is an outcome of big data analysis that allows one to not just control these environments but also to predict them and align them with corporate interests: "the point is not only to . . . interpret the world, but more importantly to acquire the capability to ultimately change it" (Mitrasinovic, 2006, p. 274). This social engineering is no small feat. It requires deep sensitivities to specific cultural forms and media experiences that can be effectively scaled and indigenized and then adopted by a vast global populace. The organizing thereby allows for numerous themes, niche orientation, customization of needs, and manipulation of desires to achieve the total experience of immersion in this highly choreographed topography. This is an ideal example of what Lefebvre calls the shift from product to productive activity following normative expectations.

One of the most powerful manipulative variables to achieve this productive activity is that of affect. Walt Disney, one of the key pioneers of the experience economy, was well aware that it is not products that require personalization but rather the experience. This has been widely adopted in the digital age as sophisticated target marketing based on sentiment analytics is put to task with every click of the mouse (Ayres, 2007). Today, the customer is lured, seduced, and convinced to enter the portals of virtual worlds and multiuser game environments. To do this effectively, customers are viewed as niche audiences and gaming environments are balkanized to reflect users' wide-ranging tastes and aspirations. Brands are very much webbed into this narrative and with them come time-tested mechanisms to foster emotional engagement within these fantasy spatial formations. Jenkins (2006) has described this catering to emotion as a new era of 'affective economics' in his book *Convergence Culture*, connoting a more positive interpretation of this. However, Mark Andrejevic (2011) lends a more critical stance to the leveraging of emotion. He emphasizes user manipulation as part of the larger capitalist system of emotional currency.

While buzzwords such as 'emotional capital' have been used time and again to celebrate the building of affect towards virtual leisure worlds, Andrejevic views this as a spectrum of emotion that can move in several and possibly unintended directions. So the question remains: Is tapping into affect to sustain users in digital and material fantasy parks devious? Or, on the contrary, is it the marketing industry's benign public service to serve their demanding clientele?

Marketing companies obviously defend themselves by framing their activities as that of servitude towards the customer:

> The refrain of the marketing industry (at least for public consumption) is that advertising does not instill desires, emotions, anxieties, but merely taps into already existing, perhaps latent, ones. If someone is moved by a targeted campaign to make a purchase that wouldn't have been made in the absence of the ad, the marketers have merely helped a consumer to realize his or her desire. This is the apparent indeterminacy of consumer desire: on the one hand reliant upon the ministrations of marketers, and on the other, an un-coerced invocation of latent subjective autonomy. Even as advertisers work to gather more information about consumers in order to manage their responses, they refer to their own increasingly slavish devotion to the whims of their targets . . . Once focused upon the 'cold' demographic facts—background, behaviour, income, etc. and eventually supplementing these with research on values, attitudes and beliefs, interactive marketers are turning to the measurement of sentiment, opinion and emotion on an unprecedented scale: that of the Internet. (Andrejevic, 2011, pp. 609–610)

Whatever the justification given, it does not take away from the fact that we are now far from the ethos of counterculture. This ethos saw participatory cultures providing untainted and fresh opportunities to create meaningful and non-corporatized public play domains within virtual worlds. While media giants continue to dominate in carving out predictive fantasies, what has changed is the rise of new actors in this digital game. Information experts and data companies are rising in influence. Companies such as Jodange are able to process mass opinions and visualize data at an unprecedented scale. Scout Labs is able to track online search buzz to detect social media sentiments and trends not just locally but internationally. These are some examples of new intermediaries in this usually tightly controlled and mediated fantasyscape. In fact, there are new applications that capture the feelings of the web via a 'vibology meter' and proclaim the directions and nature of emotional waves.

Another variable that is manipulated to entrench users in these virtual worlds is the simulation of realism. If we are to look at theme parks, the build-up of fantasy is enhanced when the environment looks more 'real.' The strategic simulation of reality allows immersion into the fantasy aspect of the setting. Corporations leverage this immersion to plug in their brands at points of believability: reality is infused with billboards and brand products. Similarly, digital games are often dotted with a variety of brands through product placements. Studies reveal that gamers allow for brand presence if it serves to enhance the realism of the game, making it a more immersive experience (Galician, 2013). After all, marketing is storytelling with a purpose. As long as the brands are true to the narrative, they gain legitimacy with the gamer. What is most important to the audience is the authenticity of the experience. However, the line is often thin between realism for fantasy and a feeling of commercial takeover by companies. This balancing act is no small feat and the rewards are mighty high. About a decade ago, game companies started to really get a sense of the gains behind such an effort. For instance, in 2004, Electronic Arts Inc. (EA), the world's largest video game publisher of the time, made about 1.5 million dollars in revenue from five in-game placement deals with advertisers AutoZone, Dodge, Levi's, Mr. Clean, and Old Spice in the game NASCAR 2005: Chase for the Cup. Net revenues for the gaming companies reached as much as 1.8 billion dollars in 2010. Today, even simple social media games for mobiles are striving to enhance their revenue by employing similar tactics. Blake Commagere, a pioneer of early social games such as Vampires, founded a start-up in California called MediaSpike that allows game developers to easily place products into the game environment in much the same way that television and film shows have in the last decades (Takahashi, 2012). In fact, product placement has become so commonplace in the mediascape that Hudson and Hudson argue that we need to reframe it as 'branded entertainment' because:

> a common theme of these more recent definitions is the term 'integration.' Branded entertainment is defined as the integration of advertising into entertainment content, whereby brands are embedded into storylines of a film, television program, or other entertainment medium. This involves co-creation and collaboration between [among] entertainment, media, and brands. (2006, p. 492)

While indeed 'Disneyfication' is seen as synonymous with corporatization of the fantasyscape, there is another equally important characteristic of this unique leisure domain that made it so beloved and endearing to the public. Sharon Zukin argues, in her book *Landscapes of Power: From Detroit to*

Disney World (1991), that while these terrains are a global spread, the message was one of American locality.

> While Walt Disney won fame as a founder of Hollywood's animation industry, his real genius was to transform an old form of collective entertainment—the amusement park—into a landscape of power. All his life Disney wanted to create his own amusement park. But to construct this playground, he wanted no mere thrill rides or country fair: he wanted to project the vernacular of the American small town as an image of social harmony. (pp. 221–222)

The exporting of this notion of American purity and of timeless innocence is packaged in a sanitized and thoroughly efficient design frame. To do so, these topographies have to be stripped of any hint of urban ills and must instead market an ideal of a democratic public realm that simulates equality among all. This is 'Disney realism,' as John Hench, its legendary park designer used to say (Hench & van Pelt, 2003), where "we carefully programme out all the negative, unwanted elements and programme in the positive elements" (p. 118). While this was seen as a major feat among architects for decades to come, others attribute the modern stultification of urban landscapes to the emulation of this Disneyesque quality. From the gaudy Las Vegas to the fantastical Times Square to the outlandish Dubai, it seems as though certain cities resemble theme parks. Fantasy worlds can be viewed as seeping out into reality and perhaps contaminating the social diversity of urban life.

> In an effort to create a place marketable to mainstream tourists and corporate tenants, a coalition of public and private elites imposed a Disney model of controlled, themed public space on an area of remarkable, if unsettling, diversity. In doing so, they sacrificed the provocative, raw energy produced by the friction of different social groups in close interaction for the stultifying hum of a smoothly functioning machine for commercial consumption. (Reichl, 1999, p. 179)

We hear a similar critique regarding virtual worlds, where the loss of diversity of expression and creativity is blamed on corporate interests. We read time and again how video games are penetrating the real world and are emotionally stunting the youth with their limited exposure to real-world enactments. We are alerted to strange incidents in virtual worlds like Second Life, where virtual marriages are cemented with real money and have implications in offline legal and social systems. It seems that predictive and simplistic spaces have their appeal. Hence, where such concerns pervade, the Disneyesque discourse can be transferred from the material to the digital domain.

Figure 6.2 A meeting in the 'public sphere' of Second Life.
Source: Hildekd/Wikimedia Commons/Public Domain.

While this may be so, there are those that defend these spheres, arguing that they have a place in our spectrum of experience: one can have deeper engagements even in the most peripheral of fantasy worlds. After all, just because a context is synthetic and superficial, it does not necessitate that meaningful social relationships and communication cannot be accomplished within these landscapes. Richard Bartle (2010) insists that while these territories are designed with limited 'emotional bandwidth,' they continue to expand our affective capabilities through their novel environments. Social rituals are performed and social capital continues to mushroom within these synthetic worlds, underlining the flexibility of our human capabilities. In fact, these artificial worlds may even solidify our faith in human nature such ephemeral and play-affirming escapism. This is sometimes an essential contrast to the mundane nature of our quotidian lives. In *Coming of Age in Second Life: An Anthropologist Explores the Virtually Human*, Tom Boellstorff (2010) takes this line of thinking further by unapologetically emphasizing the embedded virtual in our real existence. He argues this is part of human nature:

Second Life is profoundly human. It is not only that virtual worlds borrow assumptions from real life; virtual worlds show us how, under our very noses, our 'real' lives have been virtual all along. It is in being virtual that we are human: since it is 'human' nature to experience life through the prism of culture, human being has always been virtual being. Culture is our 'killer app:' we are virtually human. (p. 5)

Interestingly, today the branding of clean virtue through Disney's highly organized fantasy space no longer suffices. This is because citizens are becoming more socially conscious consumers and are exercising their consumption through micro modes of activism. To maintain its 'wholesome' and 'clean' reputation, Disney has an elaborate corporate social responsibility plan to enforce its brand. In fact, they have strong policies regarding employee benefits, community building, and environmental issues, earning them awards in their CSR activities (Clavé, 2007). For example, Walt Disney was identified as the top company in CSR by the Boston College Center for Corporate Citizenship and Reputation Institute. In 2008, for the fifth year in a row, Anheuser-Busch was ranked first for social responsibility in *Fortune* magazine's 'America's Most Admired Companies' and 'Global Most Admired Companies' lists. Likewise, we already witness how virtual worlds and gaming apps—such as in Second Life and World of Warcraft—promise educational and community-building benefits. They recognize the importance of branding these virtual domains as places that users would like to inhabit longer, and within them, enrich their social lives.

Amusements on the Go: From Flâneur to Phoneur

In Chapter 2, we touched upon the notion of the digital wanderer, the digital flâneur that serves as a powerful metaphor for modernity, urbanity, and consumption. Baudelaire characterized the flâneur as one who participates in the makings of the cityscape and yet detaches from the surrounding environment. This constant tension between the spatial and the social movement has been leveraged as a theoretical frame through which we can understand new leisurely explorations through commercial avenues. Walter Benjamin (1983) builds on this uninvolved yet perceptive attitude of the flâneur: He immerses himself in the modern-day spectacle around him and revels in the shopping arcades of the time. Benjamin explicates that even the people around the flâneur become part of the arcade experience, lending flavor to the sensory experience while not cultivating meaningful contact. If we step back from this description for a moment, we can see how the flâneur fits the description of the modern-day tourist who engages and yet disengages with fantasy parks. Always on the margins, this tourist is constructing and collecting memories that will be encased in the ideal. Theme parks enable this architecting of experience as they provide the earlier mentioned utopic realism.

The question in this digital age is about the flâneur's anonymity. Baudelaire and Benjamin emphasized this invisibility in the arcades and spectacles of consumption, of being lost in the maze of fantastical experience. Today, however, the digital flâneur is being followed closely and constantly, is monitored and measured, and every act of temporal browsing becomes a permanent consumer history. While these purposeless meanderings continue online, they become a text within a larger narrative of consumption patterns that aid and abet the corporation and the state. While the first generation of observers of digital flâneuring celebrated this freedom of movement on the web, Castells (1996) was quick to point out that this rambling has been largely forsaken for purposeful uses: "the great strange space has morphed into the space of flows, where capital seeks directed outcomes and the privileged pursue the rewards of the information society" (p. 410). The learning and technology scholar Sherman Young has alerted us to a significant shift in the strolling acts of today where exploration and discovery have given way to efficiency and deep mediation by the brand.

> When people do walk in the city, flanerie has been displaced by the precise itinerary of the three day tourist, the three minute dash from train platform to bus stop, the insistent battle between jaywalker and bicycle courier. Instead, the driveur, encouraged to incessant labour by the insistent demands of email and cellular phone; directed towards happiness by wealth acquisition and rewarded by the promise of early retirement, is the dominant actor in the twenty first century metropolis. (Young, 2005, p. 26)

Today, digital architects like programmers and game developers have discarded the romantic ideal of flâneuring. Urban planners have also cast aside the ideal when designing material fantasy arcades. Rigid zoning and constraints on time demand a preconceived plan of action by the contemporary flâneur as he labors away at accumulating all possible experiences within these manufactured landscapes. Here, the flâneur superficially absorbs the park sites around him, getting a sampling of multiple fantasies. The routine of random browsing of online leisure spaces, the making of connections through hyperlinks and of following paths and trails that lead to the unknown may be coming to an end. Joseph Turow (2005), one of the leading scholars on consumer surveillance online, reiterates this point when examining audiences who are being tracked on the web. He states that they are no longer considered as people but as constructs that feed into abstract data categories created by information specialists. He talks about the window-shopping realities of today's flâneuring: searching is not surfing, and where precision is most valued.

The study of audience movement is further complicated by new affordances of mobile phones and new apps that, via virtual navigations, guide people through the physical terrain. In fact, urban spaces are increasingly playing a role in navigation within virtual worlds, allowing the contemporary flâneur to be both digitally and physically bound. However, this seamless strolling through the wireless and physical world comes at a price. Robert Luke, the director of Applied Research and Innovation at George Brown College, argues that today's postmodern version of the flâneur is the phoneur, who lives through his mobile phone and commercially engages with his surroundings:

> Our wily e-urbanite emerges from his lair and makes his way to the local coffee shop. On the way, he pre-orders his latte with his cell phone. The venti-sized beverage with a sprinkle of chocolate is waiting for him when he arrives. Payment is automatically processed from his m-commerce phone, scanned as he walks in the door. (2006, p. 189)

Here, articulations come with bar codes within virtual corporate grids, making the phoneur's reality that of heightened conspicuous consumption. As Luke continues, "an identity is mobilized as the phoneur wanders, observed while (s)talking the city streets," all the while being "stalked by corporate hunters" who "place the social relations of phonerie amidst flows of commodity and desire" (p. 191). Today, most theme parks have apps that allow their customers to anticipate which routes to take and which sights to see in the seemingly endless array of immersive opportunities. This appears to push spontaneity and idleness further away from contemporary social practice. Likewise, mobile phone games carve out a hybrid-reality that is at once location-based and urban and yet virtual.

This is not to say that deeply structured environments cannot be playful. In fact, scholars such as Silva and Hjorth (2009) demonstrate that urban spaces have tremendous potential to be playful, and this hybridity of the virtual and physical can generate much amusement. To support the idea of playful space, Lefebvre (1991, p. 38) again lends a helping hand. Here, social space is composed of perceptions of space (perceived spaces), representations of space (conceived spaces), and representational spaces (lived spaces). Hence, playful spaces are embedded between the representational spaces of daily routines and the urban realities of leisure and work networks. After all, play is a combination of factors: the boundaries of ordinary life, of total immersion, of freedom of movement, and of paying heed to the rules that govern these spaces. Hence, we do not need to compartmentalize fantasy and play from daily life. We need not embrace Robert Luke's damning vision of postmodernity where mobile users are instruments of m-commerce and e-surveillance. There is another more optimistic vision offered by Silva and Hjorth that acknowledges the agency of the audience. Mobile games, they argue, can disrupt and subvert the normalized terrain and the mundane routine and infuse spontaneity in the usual scripts of urban performance.

Specifically, there is a genre of games called the 'Big Games' that strive to merge reality and fantasy and the urban and themed landscape. These games treat reality as mediation. Frank Lantz, director of New York University's Game Center, describes them as "large-scale, real-world games that occupy urban streets and other public spaces and combine the richness, complexity, and procedural depth of digital media with physical activity and face-to face social interaction" (Ruberg, 2006, p.2). Here, the Big Game can transform the city into "the world's largest board game," where users play detective and investigate the streets for hidden virtual treasure. Take, for instance, the B.U.G (Big Urban Game) that is designed as a five-day event across the streets of Minneapolis and Saint Paul. Through gaming, the city becomes mysterious and is made unfamiliar. Routines are broken and users rediscover their urban environment, re-creating this as a fantasyscape. Or perhaps, Shoot Me If You Can, a South Korean innovation that serves as a chasing game involving camera phones and messaging. The aim is to 'shoot' the opposing team through the lens of the camera without being shot yourself. Observers may see players physically running around downtown Seoul, preparing for unexpected encounters in the maze of the city life. In other words, this game highlights that:

> in an age of immediacy, processes of delay (both intentional and unintentional) are inherent factors. The result was a game in which *both* immediacy and delay were part of the experience, with unexpected moments like "waiting for immediacy" becoming the poetics of delay. This frustration surrounding technological lag and desires of instantaneity has often played an important part in the gameplay of urban games, and many projects have incorporated this issue as part of the gameplay strategy. (Hjorth, 2011, p.17)

Here, the mundane becomes the exotic if situated within game play. Technological affordances such as these emphasize the hybridity of real-virtual space and the hyperrealities of our time.

To conclude, such examples illustrate the importance of the social context at hand when examining fantasyscapes. In this chapter, we investigate the range of architectures and structures, as well as corporate manipulations and rules of engagement in the makings of fantasy landscapes for public consumption. Disneyland, for instance, comes with a rich literature from the field of sociology, urban planning, and cultural studies that, as demonstrated here, is transferable and applicable to digital play spaces. The metaphor of 'fantasy parks' reminds us that digital flâneuring, online marketing and branding, and the commercialism of gaming platforms and virtual worlds are in many ways rooted in theme parks of the past. This analogical tool enables us to conceptualize the emotional landscape and affective economy that pervades both the real and the virtual, and the contemporary and the historical to construct a complex interplay on the notion of fantasy.

7 Global Cities, Global Parks
Globalizing of Virtual Leisure Networks

In the end to talk about parks is to talk about the city as much as about what landscape architecture is, and what landscape architects can do.

> Jusuck Koh and Anemone Beck, 'Parks, People and City'

Conceptualizing digitization and globalization . . . creates operational and rhetorical openings for recognizing the ongoing importance of the material world even in the case of some of the most dematerialized activities.

> Saskia Sassen, *Public interventions*

Diasporic public spheres, diverse among themselves, are the crucibles of a postnational political order. The engines of their discourse are mass media (both interactive and expressive) and the movement of refugees, activists, students, and laborers. It may well be that the emergent postnational order proves not to be a system of homogenous units (as with the current system of nation-states) but a system based on relations between heterogonous units (some social movements, some interest groups, some professional bodies, some nongovernmental organizations, some armed constabularies, some judicial bodies).

> Arjun Appadurai, *Modernity at Large*

Where does the park end and the city begin? Can we talk about the park without relating it to the city? While the park is centered within the cityscape as its social and leisure public domain, it is also decentered from the dominant functional ethos of urbanity. The architects Jusuck Koh and Anemone Beck alert us to the direction that contemporary parks are taking: They are becoming more distanced and cosmopolitan, and conditioned for peripheral consumption by the passersby. What seems to diminish is the nurturing, local, and intimate design of the urban park that evokes sensuality, belongingness, and a sense of community. In making their case for a more vibrant ecology of public leisure space, they suggest a dismantling of conventional boundaries between the park and the city:

> As much as possible a park should not be bounded or bordered in a zone defined by city planners or a social sector. It must be open: visually, socially and ecologically. It also needs to be programmatically open to change, open to participation of community, open to aesthetic participation of users by using comprehensible formal languages, and open to momentary or time-share ownership of the users. Desirably, an urban park today could reach out into the city like an octopus. Likewise, it could let the city come in with its urban uses and activities, with restaurants, theaters, museums, or even complementary housing. The result would be a 'park in the city' or a 'city in the park,' realizing necessary interpenetration and mutual complementarity between nature and culture, and park and city. (2006, p. 16)

While indeed the intermingling of these two spheres can be effective in shaping a more livable social environment, boundaries continue, underlining the historical persistence of social practice within these realms. When these borders blur, we need to pay attention to points of convergence and divergence.

In prior chapters, we have seen the coming of age of urban parks. Particularly in the 19th century, this was a response to the fast-growing industrialization of the time. During this period, the state, be it China, the United States, or England, seemed to share a social vision of the urban park as a spatial strategy to foster modern civility and communal feeling. We have witnessed the omnipresence of park formations in dialogue with democracy and urbanization, two globally sweeping phenomenon that signaled the rise of the modern society. We have witnessed the park as a radical act of carving the city into an open space marked for the public, to exercise their range of leisure expressions and social enactments. Urban parks emerged as a symbolic, political, and ideological landscape worldwide.

One can argue that it is impossible to experience the park without the larger experience of the inhabited city. While distinct cultural practices mark these topographies, the park and the city undoubtedly share certain architectures and social infrastructures. They are both in constant play with forces that demand control. Both spheres are subject to the practicalities of design for standardization and uniformity. At the same time, social inhabitation of these spaces compel plurality and creativity as the individual and institutional needs strive to leave their mark on these landscapes.

This book has explicitly drawn parallels between the urban park and social network sites to highlight the privatization, commercialization, and politicization of public leisure space. In this chapter, we embark on two missions: the first is to situate the urban park as part of the larger cityscape, and the second is to underline its global implications. In parallel fashion, we frame social networking sites as part of the larger Internet domain and we underline the globalizing of the digital leisure commons. This is not as ambitious as it seems. Over the decades, the relationship between the digital commons and the material commons has matured, catalyzed by the metaphor. We have learned to conceptualize the Internet through analogies to grasp its information highways, networks, the underlying logic dictating movement, and nodes of concentrated social action (as illustrated in Chapter 2). William J. Mitchell's influential book *City of Bits: Space, Place, and the Infobahn* (1996) laid a solid foundation for comparing the Internet to the city. His prophetic perspective on the web as 'soft cities' highlight the underlying infrastructures and architectures of digital space. Using a historical and urban approach, he calls for a novel way of framing these techno-social domains:

> In a world of ubiquitous computation and telecommunication, electronic augmented bodies, postinfobahn architecture, and big time bit business, the very idea of the city is challenged and must eventually be reconceived. Computer networks become as fundamental to urban life as street systems. Memory and screen space become valuable, sought after sorts of real-estate. Much of the economic, social, political and cultural action shifts into cyberspace. As a result, familiar urban design issues are up for radical reformulation. (p. 107)

Mitchell emphasizes that while digital space appears infinite and freely accessible, it is subject to accessibility constraints and regulatory factors:

> If the value of real estate in the traditional urban fabric is determined by location, location, location (as property pundits never tire of repeating), then the value of a network connection is determined by bandwidth, bandwidth, bandwidth. Accessibility is redefined. (p. 17)

Lawrence Lessig took this comparison further in his book *Code and Other Laws of Cyberspace* (1999). He provokes us with a spectrum of ramifications: The impact could be tremendous if the code—the building blocks of these virtual architectures—can be controlled and regulated by interests that are not necessarily democratic. He warns us that "we must consider the politics of the architecture of the life there" (p. 293). Today's climate is in line with his predictions. Now, we hear talk of big data structures attempting to cement our digital experiences into predictive molds. We learn of walled gardens enveloping our leisure experiences with strong boundaries, and algorithms influencing our navigation into specific pathways. By equating the Internet to the city, we can benefit from applying our understandings of urban planning and cultural geography to current conversations on the shaping of the digital commons.

Another much talked about dimension of this realm is its network potential, where dense sociality is organized in a multiplicity of ways. Manual

Castells' book, *The Rise of the Network Society* (1996), is credited with significantly shaping the global cities scholarship. Castells convincingly maps these ideas to understand contemporary space. He argues that cities should be viewed not as places but as processes, where ideas, goods, and people flow through them. This contributes to the rich array of relations that attract us to these domains. He introduces the term 'spaces of flows' to break the conventional notion that cities are bounded entities. Instead, he shows us how communication technologies and transportation networks enable cities to be more fluid.

The persistence of this parallel has matured: We have moved from the initial utopic notion of the web as a novel frontier of limitless and depoliticized Western space (Barlow, 1996) to a more architected and socioeconomic phenomenon of a propertied and contextual digital place. The Internet realm has tremendously benefited from scholarly insights on the material sphere to aid in the architecting and conceptualizing of virtual social practice. Interestingly, the 'city' itself, while lending itself as a tool to illustrate the digital sphere, is in fact going through its own metamorphosis. There is no one generic understanding of the city. In fact, within the host of cities to choose from, we recognize persistent hierarchies, networks, and clusters that resemble the core–periphery binary. Certain cities have become templates to emulate, termed as 'global cities' (Sassen, 2001). These select cities are seen as command centers that serve as a fulcrum for the industrial, the creative, the leisurely, and the privileged as well as for temporal laborers and the migrant class. Similarly, not all social networking sites share the same power and influence. For instance, Facebook and Twitter are the virtual command centers of the digital age. Much like the global city, they are at once stateless and yet constrained by diverse national laws, local sociocultural politics and practices.

This chapter draws heavily from literature on the global cities, using it as a discursive tool to deepen the analysis. The metaphorical parallel of the city as the Internet is used as a point of departure. This makes sense as the metaphor has been established for over a decade. This allows us to use the framing of 'global cities' to capture the globalization of digital architectures. We build on this rubric to delve into a segment of the Internet that is marked for leisure—that of social networking sites. Taking a cue from global cities, this chapter reveals the globalizing of digital leisure networks through the spatial metaphor of 'global parks.'

Throughout this book, there has been an emphasis on how various urban parks reflect dimensions of social networking sites, sharing the rhetoric of being democratic, participatory, open, and leisure-oriented. While seemingly innocuous, we have seen how urban parks have a contentious history in becoming a public and democratic space. We have seen how their universal and cross-cultural attributes hold across different nations. Parks share a history of struggle in making and sustaining their spaces as public. We see that behind the design of urban leisure spaces are intentions, regulations, and constraints that are often played with by people that inhabit these spaces. By investigating the globalization of the Internet through the lens of the 'global city,' we can push our understandings on the commonality among parks across contexts. Overall, this effort allows us to do several things: First, we can constructively borrow from the field of urban studies. Second, we can extend globalization debates from the material domain to the Internet. This allows us to better confront the political, sociocultural, and economic dimensions of globalization and their online and offline intermediations.

THE CITY AND THE PARK

In the 19th century, the Industrial Revolution brought about massive urbanization, as we have addressed in earlier chapters. It promised tremendous economic prosperity and yet threatened the quality of life. The strapped infrastructures of the city designed for productivity left little breathing room for diverse public expression. Public parks were seen as a solution, a safety valve. No society can sustain itself on the purely pragmatic. People's needs, desires, aspirations, and expressions are fundamental to a lived space. Sustainability and regeneration became the underlying premises for the urban park movement (Woudstra & Fieldhouse, 2000).

Thus, a symbiotic relationship was born, where the city is nurtured by the presence of its parks and the parks cater to the unmet needs that emerge within the city. Since then, expectations of the urban park have risen as scholars have pointed out the linkages of these domains to economic, psychological, and social prosperity. It pays to have leisure. It is productive to recreate. The modern society comes with a social vision and strives to embed these values through the aesthetics of park design. For instance, equal access to public goods was a new social value. This value departed from the past practice of urban parks being accessible for only a select and privileged population. During this time, the future of the community was driven by a vision for democracy. Much emphasis was now being placed on bringing a diverse group together in these new green commons. The aim was to create connections and shared interests, contributing to the makings of a responsible and socially invested citizen of the city.

In the makings of the public greens, the municipality was spearheading this process. Soon enough, however, in the name of democracy, a range of actors began to play their part. For instance, in the early 20th century the Department of Public Works of Amsterdam, led by the architect Cornelis Van Eesteren, in collaboration with the landscape architect Jacopa Mulder, designed Amsterdamse Bos located in Amsterdam, the Netherlands. The design team was multidisciplinary, consisting of teachers, botanists, biologists, engineers, architects, sociologists, and town planners (Loures, Santos & Panagopoulos, 2007). This team involvement was useful in conceptualizing the park to meet the needs of the modern city and serve as a guardian for its sustainability. In analyzing the urban park models of the time—such as the Parc André-Citroën in Paris and the City Park of Porto in Portugal—much can be attributed to citizen involvement in the design of public space. Also, the location of the park, once on the periphery, now became central to city design, occupying prime space and defining the character of the city itself. The economic sustainability of the park led to the growing influence of corporate magnates who donated resources. The industry was quick to discover that these seemingly innocuous parks led to a rise in income and prosperity, impacting the real estate around them and enhancing the value of the city. Appeal caught on as we see the urban park movement spread worldwide during this time. The Victoria Park in London and the Central Park in New York served as templates for enlightened park design across nations, merging design with the city's best interest.

It is easy to get swept away in the romance with the public park and accept its design and architecture as normative and inherently positive for social order. After all, who hates parks in their neighborhood? Who would dispute a public good such as this for their social well-being? Yet, as we should know by now, no domain is completely sacrosanct. Jane Jacobs, a grassroots American activist of the 1960s was seen as an unlikely candidate to influence urban planning and renewal. And yet her book, *The Life and Death of Great American Cities* (1961), was seen as a pioneering work: It offered concepts that we take for granted today, such as social capital. She watched closely how the planning establishment would mindlessly imitate and transpose models for development onto existing public spheres, without paying heed to the specific social context. She enjoyed provoking the intelligentsia on how urban 'renewal,' in spite of its futuristic promise, served to create slums. She did not hesitate in questioning that which was faithfully revered at the time—that parks are good and crowding bad. She argued that at times, parks could be dangerous because of their isolation and crowded areas could be the safest inhabitable spaces.

When we shift our attention to the digital domain, we find a similar symbiotic relationship. We see that through the Internet and social networking sites, or what O'Reilly terms 'participatory architectures.' The Internet refers to a global system of interconnected IP networks, datagram structures that enable the exchange and flow of information across destinations. In other words, the Internet is marked for ecommerce, egovernance and a host of digital leisure practices. Much Internet growth and usage in the recent decade has been attributed to the popularity of social networking sites for entertainment, play and pleasure. In fact, the terms 'Internet' and 'Web 2.0' are often used interchangeably today. This signals the dominance of leisure platforms such as Facebook, Twitter, and Cyworld, as well as a multitude of media sharing sites, in shaping the perception of the digital commons. The user-generated and participatory culture appears to be seeping outside these boundaries and into most digital spheres. It is also clear that these leisure domains are viable and lucrative, propelling several private-sector actors and agencies to adopt these architectures for their public outreach. And, as Jane Jacobs reminds us, these public spheres, in spite of their participatory lure, can also be non-conducive to society. For instance, criminal activity and sexual deviance capitalize on these transnational networks to spread and garner support from a diasporic public sphere.

In Chapter 3, we discussed how virtual political activism gains global attention through social networking sites and often spills over to other digital and material domains. In Chapter 4, we discussed at length the international trend of privatization of digital leisure architectures as corporate interests enmesh with public values. In Chapter 5, we addressed the circulation of digital labor across borders, simulated as corporate leisure. And in Chapter 6, we investigated the multinational branding of gaming topographies that mark our amusements today. While the transnationalism of digital leisure networks have been addressed in prior chapters, what is yet to be discussed are the makings of command centers—both digital and material—within this highly competitive landscape.

GLOBALIZATION OF THE URBAN AND THE DIGITAL COMMONS

The Global City, Command Centers, and Corporate Networks

Cities are not equal. Understanding their hierarchies and the extent of their global reach has been approached in the last few decades through the construct of 'global cities.' This concept's well-known proponents, John Friedmann, Saskia Sassen, and Peter Taylor, argue that because of economic, political, and cultural factors, certain cities are disembedded from their national systems as they exert their presence in global ways. Contrary to the popular notion that globalization is impervious to borders (Held & McGrew, 2000; Friedman, 2006), Sassen's innovative argument on the 'global city' is built on the emphasis of boundaries: The city's unique centrifugal localization contributes to the denationalizing of these structures. In other words, New York, Tokyo, London and Paris exist in a bubble zone of particular politics. This is attributable to their unique capacity to attract and sustain a global flow of sociocultural and economic capital. Thus, these cities hardly mirror the larger national culture within which they are situated.

Sassen defined the term 'global cities' in her classic book, *The Global City* (2001), as "strategic sites in the global economy because of their concentration of command functions and high-level producer-service firms oriented to world markets; more generally, cities with high levels of internationalization in their economy and in their broader social structure" (p. 154). In reviewing the burgeoning literature on this term, the following main characteristics come to light, as Brenner and Keil frame it (2006, p. 11), namely:

- basing points for the global operations of transnational corporations
- production sites and markets for producer and financial services
- articulating nodes within a broader hierarchy of cities stratified according to their differential modes of integration into the world economy
- dominant locational centers within large-scale regional economies or urban fields

In conceptualizing the 'global city,' much weight is given to financial markets in the reorganizing of cities' spatial structures and, within those, a new transnational class system. Sassen (2006b) paints the picture of such cities as "command points" (p. 2) of corporate power that foster a formation of networks. These networks are not just among the financial elite but also among the low-paid immigrant service workers that sustain these city economies.

William Carroll (2007) reinforces this linkage between global cities and transnational corporate networks. In his study, he argues for the staying power of nationhood as these entities continue to be constrained by regional legalities, politics, and economic underpinnings. It is worth remembering that corporate power is not a consensual and centralized force. Rather, it comes with its own internal tensions that often are reflective of specific local/national conditions and affiliations. For instance, we cannot assume that Shell Corporation in the United Kingdom is seamlessly operating along the same lines as Shell in the Netherlands. These organizational entities are more segmented in form through their strategic, operational, and allocative features:

> Strategic power occurs at the level of structural decision-making and concerns the determination of basic long-term goals and the adoption of initiatives to realize those goals. Operational power involves the actual implementation of corporate strategy within the head office and in sub-ordinate offices, subsidiaries, and plants. Finally, there is the allocative power wielded by financial institutions, whose collective control over the availability of capital gives them the power to determine the broad conditions under which other enterprises must decide their corporate strategies. (Scott, 1997, p. 139)

Thus, multiple national actors and processes play a role in creating a multinational corporation. They also exert influence in the formation and the enactments within a global city. While the state continues to exercise power, in this era of global outsourcing, workers from distant nations can and do exercise their voice, even when not situated in the West's command centers. Continuing with the example of Shell, the Netherlands branch was hardly impervious to the pressures that emanated from the Nigerian Kula community in the Rivers State: They demanded the corporation keep their promise on sustainable development. Or, for instance, consider the disastrous and deadly collapse of the garment factory in Dhaka. This tragedy touched a nerve with apparel consumers in global cities such as Berlin, Dublin, and Helsinki, pushing the multinational garment industry to respond to this situation. Part of this responsive network has to do with the mobility and circulation of labor and more sophisticated communication technologies that allow for interactivity, engagement, and public awareness of local issues.

Such examples can serve as a sign of optimism against the normative hegemonic structures that have dictated these formations historically. Yet Taylor (2004) points us to the more disturbing potential of these interurban configurations: the making of a 'new network bourgeoisie'—a global plutocracy. In other words, global cities and transnational corporations strategically cooperate to sustain their overarching power, creating an urban fabric that is more impenetrable to lesser cities and agencies. In fact, the global city concept has come under severe criticism because of its bias towards the West, negating the rise and influence of cities in the global South. While cities such as New York, Tokyo, Paris, Amsterdam, and Berlin get evoked time and again to illustrate their commanding roles, cities in emerging markets gain a fraction of attention. This is important to take note as the world's substantive population resides along these peripheries and are gaining stride in their influence and reach. While power continues to concentrate and circulate within the traditional Western clan of global cities, with the rise of emerging markets, there has been some flux in membership. For instance, Mexico City, once deemed a peripheral city, has moved to the core, while certain industrial centers such as Detroit have been peripheralized (Sassen, 2002a).

These recent trends remind us that global cities are not static entities. Rather, they are evolving and transitioning and transforming. It is fairer to term them as 'globalizing cities' (Marcuse & van Kempen, 2000) as it better captures the dynamism of this category. In this light, different cities compete for this status through their ongoing restructuring and reinvention. Lastly, it is important to consider the ramifications that this term can have on urban convergence and normalization. Across the globe, cities are compelled to follow this prescribed model, compromising on the range of potential diversity that city formations can display. Robinson (2002) denounces this trend by stating that "global cities have become the aspiration of many cities around the world" (p. 548). She argues that this can have particularly devastating consequences on less economically prosperous cities. Such cities are pressured to imitate these models at the price of equality in citizen participation, access, and usage. Interestingly, this has created a new spatial form of Special Economic Zones, particularly by BRIC (Brazil, Russia, India and China) nations wherein certain regions within the state gain significant privilege over others. In this new terrain the flow of services is relatively free of state regulation and interference, unlike zones outside this boundary. Hence, global cities can be deliberately architected within emerging markets to appear less nationalistic and more international. This allows states to compete in the digital and global marketplace.

Finally, another persistent 'zero-sum' juxtaposition that circulates is that the 'global city' is pitted against the state: the strength of the city comes at the price of the weakening of the state. On the contrary, the nation-state may well be behind the rise of certain global cities. This is because strategic maneuvering may help situate the state prominently on the global landscape via global cities. Privileging certain cities enables territorial domination, creating 'glocal' nodes of accumulation and regional competitive advantage (Brenner, 1998). Thereby, it is naïve to assume that the city and the state are positioned as diametrically opposed. Instead, we should seriously consider the possibility that these actors often collude politically and otherwise.

Not coincidentally, we find parallel discussions regarding the Internet and its globalizing potential. Much like a cityscape, the Internet is a techno-social infrastructure of nodes and networks. How these structures are connected and mapped on a global scale is of ongoing concern. The birth and proliferation of the Internet has been linked to the phenomenon of globalization, some believing that these new digital structures circumvent the state and create new affiliations (Graham & Marvin, 2001). Through this lens, communication networks are seen to strengthen social relations, creating a culture that clusters across borders (Rosen, Barnett & Kim, 2011). Others argue that these communication networks extend the state reach and give greater control in all social spheres to an unprecedented degree, at times in collusion with multinational corporations.

Borrowing from the global cities literature can meaningfully enhance this conversation. We need to first identify which are the digital command centers and gauge their sphere of influence. Candidates such as Facebook and Twitter come foremost to mind. These platforms are indeed appropriated by several nations and are constantly being subject to local rules, regulations, and policies. If we look at the approach on citizen's privacy, the operationalizing of these digital infrastructures differ based on whether they are in Europe or in the United States. This is attributable to specific regional policies regarding privacy. In addition, in light of the umpteen media stories circulated in the last few years on their role in the 'Arab Spring,' we recognize the platforms' ongoing negotiations with different states that attempt to control these digital spaces.

Undoubtedly, given the platforms' global outreach, their tremendous power in dictating the rules of the game is hardly debatable. Yet, because of corporate interests, they are compelled to cooperate and even yield to the interests of the state. However, take the case of China and its dominant social media platforms such as SinaWeibo, Renren, Tencent, Douban, and Wechat. They enjoy tremendous support from the state as long as they demonstrate sensitivity to political needs through self-censorship. Hence, it is a mistake to frame this conversation as a dichotomy between the state and digital platforms. Instead, it is useful to look at the complex interplay of power that circulates between these two entities. As the global cities literature explicates, at times the state proactively propels certain global cities to the forefront to achieve international recognition. This helps them meet their goal of being a major player in the transnational domain.

Interestingly, the core–periphery model that has been used to assess the membership of global cities and to gauge whether we are facing a 'new network bourgeoisie' is also applicable in discerning the globalizing of the Internet. A study recently published by Park, Barnett, and Chung (2011) unravels the relationship between globalization and the Internet. They look at changes in connectivity by comparing countries' global communication networks in 2003 and in 2009. Using network analysis, the research was carried out on the web-based network that linked the country codes of top-level domains. The results indicate that the 2009 international hyperlink network is completely interconnected. G7 countries and Spain are at the center of the network. At the periphery are the poorer countries from Africa, Asia, and Latin America.

These findings resonate with the global cities clan of command centers situated primarily in the West. This propels us to acknowledge the persistence of the digital and material plutocracy that exists in global network formations. The study strived to determine whether the Internet has become more individualized and fragmented or whether this digital space continues to function through the conduit of the nation-state. What these researchers found was the reigniting of the classic world system theory as their data fell along lines of core, periphery, and semi-periphery relations and dependencies among these states, calling to question the independence of their digital networks. It appears that a nations' development can be understood by considering "the systematic ways in which societies are linked to one another within the context of a larger network of material, capital and information exchanges" (Barnett & Park, 2005, p. 1117).

The global cities analysis gives disproportionate attention to the economic while negating the sociocultural. The analysis of the globalizing of the Internet has been accused of a similar bias. Thereby, these researchers pay heed to this critique and go beyond economic aspects. They hope to delve further into claims of decentralization, regionalism, and cultural pluralism that have been attributed to these platforms. They found that while indeed there was a global system linking these nation-states, there were also regional clusters around language, culture, and geography that circumvented conventional borders. It is clear that the world systems theory largely prevails in terms of inequality where wealthy nations are more interconnected than the less prosperous. However, since 2003, emerging markets have undoubtedly made a significant rise on the international stage. Further they function more stably and are centralized as a regional cluster more so than in the past. The research also found that while in 2003 the U.S. was central in this mapping, by 2009, it shared the stage to a substantive extent with Europe, particularly Germany.

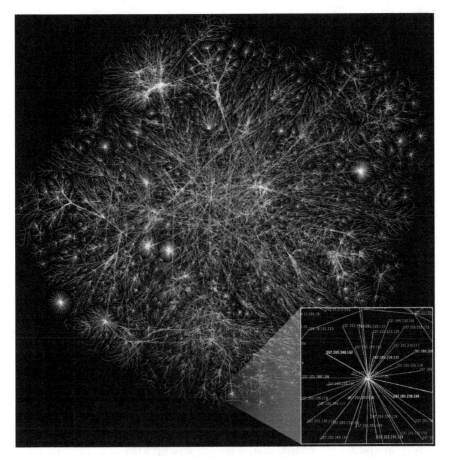

Figure 7.1 The Internet map of global information flow.
Source: The Opte Project (CC-BY-2.5 [http://creativecommons.org/licenses/by/2.5]),
via Wikimedia Commons.

Much attention has recently gone into the rise of the BRICS nations
(Brazil, Russia, India, China and South Africa), creating tremendous spec-
ulation about their role in the shaping of the Internet. Park, Barnett, and
Chung (2011) found that not all emerging markets are equal; Brazil and
Russia are more hyperlinked on the global stage than China and India.
This can partly be explained by the fact that China and India have some of
the world's largest internal digital economies. Despite China's formidable
role in today's global economy, it is far less central in the international
hyperlinked network. This has been attributed to China taking on different
language code systems to create a regional cluster of their own, a deliberate
effort to shape walled gardens within the 'Great Firewall of China.'

In fact, recent studies have emphasized the heterogeneous nature of the
Internet, adopting the term 'Internets.' The term was once a social meme

driven by President Bush's gaffe, but it is now a more serious proposition. Corporate and state politics are slowly but surely encroaching on this digital territory. The Internet was originally designed to be controlled not by any one agency, institution, or state. A case in point is the rise of Internets such as the 'Chinese Internet' with their own digital firewall, effectively filtering information flows along the line of state interests (Zhong, 2012). In fact, Lindtner and Szablewicz (2010) argue that China has not one but multiple Internets:

> the interface, content and wider social meaning of Internet technologies today are not determined by software developers and designers alone, but rather by a complex web of actors, including, but not limited to, users, corporations, state actors and policy makers. As such, it is important to acknowledge that online practice, including such things as the use of search engines or the creation and modification of digital content, is not divorced from cultural processes, e.g. social discourses and political debates. Rather than portraying the rapid changes of the IT landscape in China as a single, unified process, we stress the importance of tracing 'multiple Internets,' the development of which are contingent upon broader cultural changes such as shifts in socio-economic class, political projects of modernization and economic reforms. (p. 2)

That said, one must not forget that what has been covered in prior chapters on how Chinese citizens play with these structures to express and advocate, contribute to a far more dynamic and complex digital space than the popularly touted authoritarian perspective. As corporate, state, and other actors with vested interests contribute to the development of the Internet, it becomes clear that there is no one unified agenda and policy dictating the direction and nature of this digital geography. This is not to say that there are no hierarchical influences among these forces. Derudder *et al.* (2010) examined digital network flows and their overlap with European urban networks. They wanted to detect economic connectivity. They found that centrally located European cities had a higher influence and degree of information flow as compared to peripheral European cities with similar levels of physical connectivity. To illustrate, the combined cluster of London, Paris, Frankfurt, and Amsterdam is viewed as the 'Internet diamond': these global cities serve as important nodes of the European urban and digital commons (Tranos & Gillespie, 2011). This power fulcrum also extends their tentacles well beyond the region, marking the European influence on the global Internet backbone network. These power politics remind us that select forces shape the Internet as a public good. In spite of the affordances of communication technology to foster unprecedented locational freedom and mobility, we continue to witness the forces of agglomeration driving the exponential growth of the cityscape and the infoscape.

In pursuing this urbanization and digitization of space, perhaps it is more effective to adopt a more integrated discursive stance. This comes at a time where it is evident that global cities are becoming more mediatized and digital networks are more entrenched in their urban geographies than originally perceived. For instance, let us focus on the shift in core–periphery memberships among cities and the implications on their digital networks. Within the emerging market domain, Mexico City, currently number two, will drop to fifth place, while Mumbai is forecast to move up from third to second position. Also on the way up are urban areas such as Delhi (up from 6th to 3rd position), Dhaka (up from 10th to 4th), and Lagos (up from 14th to 7th). This indeed displays certain dynamism in periphery categories. But what is also observed is the simultaneous shift in scale and speed of Internet infrastructures and network concentrations of these geographical nodes. Recent studies have demonstrated the staying power of interconnectedness between the virtual and the material domain of social life. The Internet and city infrastructures seem to synchronize to create a complex and rich understanding of globalization of social networks and structures.

This perspective comes to fruition through state efforts to create a digital presence of 'smart cities.' This momentum rides on the convergence of online and offline structures via new technologies (Komninos, Pallot & Schaffers, 2012).

> The digital space of cities is also described as a system composed of four concentric rings. At the center are the broadband networks, wired and wireless infrastructure, and the access devices enabling communication, data collection and exchange. Then, web technologies enabling data storage, processing, and visualization constitute a second ring. The third ring is composed of digital applications in many different domains of a city for e-government, utilities management, and sustainable development. The outer ring is constituted of e-services, a few selected applications that achieve developing viable business models and offered on a regular basis as services. (Komninos, Pallot & Schaffers, 2012, p. 123)

The intense localization of these digital cities utilize a 'mirror' logic: in a sense, there are web-based representations and reproductions of different zones of the real city. This is designed to amplify the city functions and transform urban configurations to sync better with their online counterpart. Some relevant examples are the AOL digital cities that collect tourist and shopping information and couple it with local advertising. Other examples include Digital City Amsterdam, which is a platform for various community networks and social interaction among citizens, and Virtual Helsinki, representing a 3-D reconstruction of the entire city. There is also Digital City Kyoto, representing a 3-D virtual space enriched with avatars offering information related to city traffic, weather, parking, shopping and sightseeing (Ishida, 2000).

It is not just the core cities that take on a web presence. Sometimes periphery cities embrace this representation to gain entry into the core group through the back channels of the web. There is an expectation and hope that its digitization will eventually enhance its real city status. The case of Manchester is a good example: It has created a smart city rubric to become more inclusive, creative and sustainable through the imaginative use of digital applications and services. This illustrates a city's commitment to open innovation through the co-production of digital services for citizens and tourists alike to engage with their city. This comes at a time where, globally, the leisure and tourism industry is growing exponentially with the rise of the middle class. These systems of urban/digital navigation allow a simulation of inhabitation that compels one to engage in both spheres, often with positive economic and sociocultural repercussions.

At this point, there is sufficient evidence to argue for a relationship between the construction of global cities and the globalization of the Internet. We see how global cities gain a virtual presence and a digital embodiment while Internet architectures impact how we experience and architect our lived environments. This blurring of online and offline social life creates a more integrated understanding of these infrastructures and their political and socioeconomic dimensions. As a metaphor, the global city serves to represent contemporary shifts in the globalizing of the Internet. We detect a significant shift from a more generic model and rubric of the city/Internet to a more heterogeneous and decentralized model of global cities/Internets.

Recent empirical studies have revealed that there is much in common between these two constructs as we witness hierarchies and strategic networks of these command centers. Be it the physical structure of cities or the coded arena of the Internet, neither can be viewed in isolation but rather, as part of a cluster of domains that are positioned by economic and sociopolitical advantages. More than we assumed, there is mobility and dynamism within these conventional power structures as they get slowly challenged by emerging markets. This creates a diversification in network cultures, both materially and digitally. The role of the nation-state is far from disembedded within these structures. Implicitly and explicitly, and in cooperation with the private sector, the state creates walled gardens based on unique arrangements and negotiations. In fact, the state need not be diametrically opposed to the global city. Rather, it may be the primary propeller in the creation of these city entities, as it sees such action as part of an agenda to compete on an international level.

Transnational Public Spheres, the Global–Rural, and the Cultural Metropolis

Earlier we mentioned the creation of digital clusters that circumvent national borders, based on cultural commonalities like language, political alignment, and media interests. Numerous studies have explored the makings of a transnational public sphere in the age of new communication technologies. These technologies emerge in response to contemporary events and flows. For instance, the 'Islamic public sphere' addresses global alignments in the religious arena, while the 'diasporic public sphere' situates the migration of people and their sense of identity on a global stage (Fraser & Nash, 2013). Crucial to this space is a sense of inclusivity and democracy in participation. Yet such participation is not defined primarily by the state but by other rules that are more specific to the culture of these spaces. As one would expect, these spheres do not come without critique. In some sense, popular framing of the transnational public sphere has fallen into similar trappings of Jürgen Habermas's public sphere theory. Some accuse this classic Habermas perspective as being mainly informed by a Westphalian political imaginary while espousing egalitarianism. Feminists and multiculturalists and antiracists pick these notions apart: much evidence substantiates the fact that participation is rarely equal. Current power structures influence these domains to favor select groups and people. We have seen this in past chapters in the makings of the public park. There are also numerous examples of gender, race, and ethnicity bias within a range of social media platforms.

Hence, it is certainly not new to see this discursive formation of communities around aspects other than nationalism. Arjun Appadurai addressed this subject about two decades ago in his book *Modernity at Large: Cultural Dimensions of Globalization* (1996). In this work, he provides a much needed framing to tackle the nebulous notion of globalization. Still used, today, he offers spatial metaphors such as ethnoscapes, mediascapes, ideoscapes, financescapes, and technoscapes to critically construct networks of culture that are at once local and global. Particularly of relevance is his emphasis on the history and geography of a context. He emphasizes this to enable a more grounded discourse surrounding a context's cultural traffic. After all, there is no 'new' inhabited territory that is independent of the old layers of people, cultures, and things. The old and new intermingle and often reinvigorate one another. In his recent book *The Future as Cultural Fact* (2013), he pushes us further in this journey of ideas. He envelops us with a complexity of networks on what makes a global phenomenon. For instance, the diamond trade is not just situated in the global cities of London, Antwerp and New York, he argues, but is deeply connected with extreme violence in peripheral places like Sierra Leone, Congo, and Angola, and the marketing middlemen in India. Thereby, the 'production of locality' in a global city such as Antwerp is deeply affected by these global cultural flows. In this age of digital and material networks, he reminds us that we need to be more critical of our understanding of the 'flows' that these networks generate. He suggests that we "distinguish the problem of circulation from the problem of connectivity" (p. 65) when examining their levels of influence. Another illustration is the Turkish guest workers heavily circulating between Turkey and Germany and, yet, there is far less cultural connectivity between these two groups.

Keeping this in mind, how do we make sense of emerging digital platforms and applications that promise to transcend locality? Today, mobile technology apps are being designed with a global diaspora in mind. Digital audiences promise to be fragmented, fractured, and at times, fictionalized as local, while in reality they can be everywhere and everyone. Several of these participatory sites, such as Wikipedia (the global encyclopedia), Kickstarter (one of the world's largest crowd-funding sites for creative projects), and Groupon (an international discount and marketing site), appear to be digital command centers in their own right. Much like global cities, these entities feed on transnational alliances and appear to be bounded, not by the state but by their unique cultural sphere. This keeps their audiences faithful to them. In fact, if we are to illustrate this further with the example of Groupon, we find their sphere of influence wide-ranging and global, spreading across Europe, Asia, South America, and North America with 35 million registered users worldwide. While founded in Chicago, this is hardly a Chicago platform any longer. Yet if we are to examine their production of locality, they are deeply embedded in the historical practice of coupons, or discounted gift certificates used at the local supermarket or restaurant. For this digital platform to thrive and scale in diverse locations, great effort goes into creating alliances with specific local companies, municipalities and service industries within each city. Their offers reflect the local cultural needs, demands and desires and not some abstract consumption of global products and space. In fact, the offers are fundamentally local in nature while subscribing to a global template of digital discounting. Taking a cue from Appadurai's framework, the circulation may be high in a sense that audiences on the Groupon site can participate and engage at an international level if they so desire, enabled by the site's global architecture. However, as it plays out, their connectivity of consumption is often deeply localized based on old habits of frequency to nearby restaurants and their favorite food markets.

Interestingly, a new trend is emerging that recognizes the power of locality and seeks to directly cater to this traditional village-like aspiration. As urbanization becomes the signature of our time, we may find ourselves lost in the crowd. Some platforms leverage on this fundamental concern: how is it that we do not know our neighbor but we are well acquainted with an online stranger? If we have the time to fight for a global cause, why not invest in local politics? Should we not act locally if we are to globally engage? This ethnos has given birth to several web spaces such as SeeClickFix, a platform in the area of city governance that enables users to report non-emergency issues for improving the neighborhood and the city, and Localocracy, which gathers citizens, government officials and journalists to discuss and learn about local politics and priorities. The rise of the social networking site Nextdoor builds on the mission to get to know your neighbors. Their digital manifesto reads as such:

We are for neighbors

For neighborhood barbecues. For multi-family garage sales. For trick-or-treating

We're for slowing down, children at play.

We're for sharing a common hedge and an awesome babysitter.

We're for neighborhood watch. Emergency response. And for just keeping an eye out for a lost cat.

We believe waving hello to the new neighbor says, 'Welcome' better than any doormat.

We believe technology is a powerful tool for making neighborhoods stronger, safer places to call home.

We're all about online chats that lead to more clothesline chats.

We believe fences are sometimes necessary, but online privacy is always necessary.

We believe strong neighborhoods not only improve our property value, they improve

each one of our lives.

We believe that amazing things can happen by just talking with the people next door.

We are Nextdoor. We are simply you and your neighbors, together.

The burgeoning of these public spheres is seen as part of the 'slow movement,' a social response to the pressures of an accelerated society. With new communication technologies, there is deep concern that human relations are being rushed. If we want to have meaningful connectivity, we need to deliberately slowdown in our engagements and reflections of the day to day. This movement currently claims 83,000 members in 50 countries, which are organized into 800 local chapters. The World Institute of Slowness has been set up to teach the way of slowness and how it can be instilled in all walks of life. This has spawned an entire genre of slowness, including that of Slowart, Slowcity, Slowcoffee, Slowdesign, and Slowtravel. This is not a Luddite approach. Instead, there is an effort to make new media spaces work to serve one's desired culture rather than be thrust into the speed culture.

When we talk about global cities, it is worth remembering that the idea of the city emerges from the larger dichotomy of the urban–rural divide. So when we attribute the globalizing of their spaces, are we to assume that the rural landscape is situated on the outside, excluded and disassociated? On the contrary, Lise and Peter Nelson argue that the 'global rural' needs to be incorporated in this discussion. This is particularly relevant when imagining new trends of gated communities and new city constructions. These trends are targeted at an international and elite diaspora, particularly in emerging markets. They challenge our conventional perception of boundaries in a global city. These socio-spatial fragmentations are characterized as "new spaces of exclusion" (Broudehoux, 2007, p. 387), heading towards the "ghettoization by choice of the rich" (Nelson & Nelson, 2011, p. 443). Furthermore, a larger trend can be observed: Leisure spaces are permeating cityscapes to an unprecedented degree as described in Chapter 6. This is the case not just in economically prosperous regions but also in lesser states. A fast-growing consumerist culture is gaining ground within these scapes along with a unique form of spatial organization, consumer expectation, and need (Dupont, 2011).

These global rural spaces of exclusion have a significant impact on the distribution of access within the digital sphere. We have come across numerous examples in Chapters 4 and 5 that illustrate how the formation of material gated structures align with a deep digital divide. From China to India to the Middle East, the rise of special digital and economic zones is seen as endemic to urban and digital policy-making. Take, for instance, the Dubai Internet City (DIC), an information technology park created by the government of Dubai. It is structured as a free economic zone to attract and station multinational companies. It is seen as a crucial strategy for this region to become a key player in a competitive digital and material sphere. Thus, it has led many global information technology firms, such as Microsoft, IBM, Oracle Corporation, Sun Microsystems, Cisco, HP, Nokia, Cognizant, and Siemens, as well as UAE-based companies such as i-mate and Acette, to move their regional base to the DIC. The neighboring industrial clusters, such as Dubai Media City and Dubai Knowledge Village, further strengthen the zone. However, this has consequences on how the government distributes access to quality digital infrastructures among their people, privileging certain zones over others.

Part of this drive, especially among emerging markets, is to be recognized as serious players in this global arena. As these states gain a new sense of self-confidence, it becomes important for them to exercise their newfound power. It is a matter of pride that they are equal consumers and can compete within this transnational public sphere. It is not a coincidence that these states are adopting the Silicon Valley template. This comes with the embracing of 'foreign' and 'Western' emblems of status. For instance, Delhi is seen as a city that is currently vying for the position of 'global city' as it breeds a new host of shopping multiplexes. These multiplexes consist of primarily foreign retail chains offering brand-name goods and services, devoted to an exclusive and cosmopolitan clientele. These spaces symbolize an aspiration for a global culture that is implicitly foreign and Western. Basically, the global city comes with an expansive ideology that privileges certain consumption patterns over others. This preference is architected through institutional and regulatory measures such as formulating laws on what constitutes as appropriate spaces and activities of consumption that differ from state to state (Tseng, 2011). However, it is not necessarily a process of imitation:

> Although the middle classes' cultural aspirations have played a critical part in the global drive observed in India's big cities, the new

consumption patterns backing physical transformations may be part of a global modernity without necessarily being reducible to an imitation process of mere Westernization . . . In Delhi, this is best illustrated by the recently built 'hi-tech religious and nationalist theme park' devoted to a reconstructed Vedic civilization and Hindu identity, the Akshardham Temple complex, on the banks of the Yamuna river, whose conception draws on many ideas from Disneyland and Hollywood studios . . . Akshardham is analysed as the embodiment of a process of 'moral consumption' characterizing a fraction of the middle classes that 'can take part in the process of [foreign] modernity, but also 'pull back' and return to 'tradition' to preserve its 'true' Indianness . . . The anxieties engendered by the destabilizing contradictions between globalization and national identity may thus produce a 'hybridized form of globality.' (Dupont, 2011, p. 548)

This ropes us into the much-discussed 'McDonaldization' or 'Disneyfication' of the cultural landscape as illustrated in Chapter 6. This refers to concerns related to the extent of standardization, Westernization, and uniformity that globalization brings to the table. These discussions are balanced by localization and appropriation examples. For instance, scholars emphasize how even a city's 'global' cache is in actuality its local environment and culture and not some generic international lifestyle. For instance, in his study of Shanghai, Yen-Fen Tseng (2011) captured the perceptions of skilled Taiwanese immigrants who placed significant value on the unique and distinctive charms of the local environment: "skilled migrants are not as hypermobile as imagined. They value the cultural attractions and lifestyles associated with particular destinations, and are inclined to put down roots once they have settled in a new place they call home" (p. 766). This creative class is mobile and has exposure to other global cities. They share a demand—common to other skilled labor across the globe—for what constitutes as quality of place. Richard Florida states this elusive quality emanates from an assemblage of "thick labor markets, lifestyles, social interaction, and diversity of a cityscape" (2002, p. 32). Hence, the 'global' in a city can be very much indigenous in nature. Further, the staying power of supposedly hypermobile migrants are often tied to the value they place in a city's culture.

Another study by Molotch (2002) focuses on automotive designers. It determines how characteristics of place influence the designers' migration and settlement patterns. Molotch found that beyond simple job-market considerations, certain types of designers tend to prefer specific cities, because of the lifestyles, subcultures, and leisure activities that characterize each place. This brings to our attention the role that leisure and culture play in the shaping of spatial constructs. While the discourse has been dominated by economic and financial concerns, scholars such as Soja (2000) have emphasized the cultural dimension, framing the global city as a 'cultural metropolis.' After all, these spatial configurations are lived architectures where motivations are not merely driven by the pragmatic but also the affective. People seek meaning and value in a place of inhabitation; sometimes its social memory can have a powerful impact on such choices. A good example can be drawn from Tseng's (2011) analysis of how the Taiwanese feel towards Shanghai, where their "affection can be traced back to collective memories formed by representations in literature and films, such as images of old Shanghai represented in many films based on the work of famous writer and Shanghai native Eileen Chang" (p. 776).

In fact, hypermobility and digital migration are topics also discussed within the Web 2.0 sphere, as the barriers of entry are low and mobility of users high. All it takes to enter the portals of Amazon or Couchsurfing is to set up an account. We have witnessed a mass digital exodus of audiences from MySpace to Facebook. We have also seen how Google+, in spite of their draconian strategies to get people to socialize in their space, is so far failing. We are supposed to be an attention deficit and detached populace that moves with the digital herd. Particularly, the digital natives are supposed to be digital migrants at heart as they flirt with multiple platforms and apps. Yet, for the most part, this is not so. Instead, there are many examples that illustrate how people invest considerable time and energy into developing their profile, making specific digital sites their 'home.' Some audiences deeply align their identities with certain platforms, seeing themselves as Reddit people or 'Couchsurfers.' These positions come with hierarchies among members based on how often you post, your tone online, your mediation skills and how you play the rules of the game. New research has looked into audience participation on social media platforms in poverty-entrenched areas of the global South. Rangaswamy and Cutrell (2012) describe the deep aspirations youngsters have as they go about befriending strangers from exotic and Western countries. They do so to enhance their social capital on platforms such as Orkut and Facebook. Thereby, we need to recognize the nuanced differences among inhabitants of these spaces and move away from holistic groupings.

To conclude, the transnational public sphere serves as a unifying meta-narrative. It is driven by a compelling idea and propelled by the rise of new digital connectivity and circulations of people, products and ideas. Whether or not this concept had initial empirical evidence seems now less relevant. The public sphere has been reified through the concentration of resources—intellectual, financial and social, acting as a magnet for skilled labor (Martinez-Fernandez et al., 2012). Sometimes, there are dark ramifications when certain 'global cities' dictate the new economic order while lesser cities starve for capital. There is the challenge of subscribing to a normative ideal as these global structures become more sophisticated, reflecting complex needs and demands of pluralistic audiences.

We have seen that these supposed hypermobile migrants desire more rootedness and the intense localization of the inhabited space plays a large role in fostering loyalty to the domain, be it digital or material. Clusters form along lines of language, taste, and political affiliation that need not collide with nationalism. It is worth considering the application of 'globalizing cities' as a more appropriate metaphor for the globalizing of the Internet. This is because dynamism is the inherent quality that is valued in both realms. While markets play an important role in the position of these select spheres, culture continues to matter and can supersede the pragmatic.

While the classic core–periphery dichotomy has been applied to the understanding of networks, these are not essentializing categories but have more mobility than portrayed. It is popularly assumed that the state is losing relevance. But, as revealed here, it is not a simple zero-sum game of the state versus the global city. Both can be, in fact, key propellers of certain domains to reach a global stage, be it through a digital platform or a material sphere. And while the template that circulates in the architecting of these spaces can be rather uniform in nature, they hybridize as they gain usage. Lastly, the digital migrants are less uprooted than they seem, as they need that which has always shaped social groups: stability and community.

GLOBALIZATION OF DIGITAL LEISURE NETWORKS

So far, we have made an effort to tie the discourse of the global city to the globalizing of the Internet. We have done so by borrowing from discussions within urban planning and cultural geography. The intent is to create a channel of transference through the vibrant 'global cities' metaphor to facilitate discussions on the globalizing of the Internet. This section takes this further, focusing on the arena of the Internet that is demarcated primarily for social and leisure purposes—its social networking sites. These digital leisure networks have managed to carve a substantive niche and facilitate a cultural shift within the Internet sphere. Marked by social relations and leisure generated activity, Web 2.0 spaces have been attributed to create novel digital markets and online business models that come with significant economic and social value.

Yochai Benkler's *Wealth of Networks* (2006) and Henry Jenkins's *Convergence Culture* (2006) laid the foundation of tying cultural production and consumption to economic processes, with a special focus on labor-based practices. To investigate the globalizing of digital leisure spaces and the issues that pervade, it is important to situate them within the larger Internet infrastructure. As suggested at the start of this chapter, the 'city' has served as a useful metaphor to understand the Internet. It seems logical to extend this metaphorical effort to that of 'global parks' within the cityscape to address digital leisure networks.

To begin, social networking sites share much of their underlying architectural qualities with that of the Internet: they draw upon virtual and overlay networks, cloud-computing architecture, network management and traffic engineering, addressing and routing architectures, broadband access technologies, resource allocation and more (Papacharissi & Mendelson, 2011). Among these leisure sites, peer-to-peer networks and user-generated content prevail as an overarching and international template, allowing them to fit well under the larger 'global parks' construct. However, there are also distinct lexical structural differences among sites such as Friendster, MySpace, Facebook, Orkut, and Renren. These differences make the sites intensely localized in terms of their cultural space and yet globalized in terms of their legal and economic status. For instance, Stutzman (2006) tracked the types of social activity on such sites and found that across different platforms, few actors personalized their platforms by modifying their default settings, such as for privacy. Yet the behaviors that developed within such architectural confines were indeed diverse, based on the nature of communication, relations, and membership of such networks. While these sites enjoy a certain global status through their shared design rubric, certain interests such as privacy make these platforms uniquely political. These interests also tie the sites to the eccentricities of the specific nation-state within which usage happens (Gross & Acquisti, 2005).

Similarly, in the 19th century we witnessed how the urban park as an innovative public sphere was adopted across nations. With that, expectations developed of nurturing urban civility, controlling social unrest, and signaling modernity on the world stage. Yet each park comes with its own narratives and historic dramas. There is a genre of parks marked by politics, such as the People's Park in Berkeley as described in Chapter 3. There is another genre of parks marked for walled leisure consumption by the elite class, such as Gramercy Park in New York as illustrated in Chapter 4. In fact, this book has revealed a typology of urban parks based on a spectrum of cultural norms and social practice. Each genre of urban parks subscribes to a certain transnational public sphere that is not dominantly characterized by their state. However, to understand the quotidian relations that unravel in these spaces, the locality of politics and eccentricity of embedded social practice come to the fore.

Sometimes there is an overt state commitment to avoid the transnational public sphere, as it is viewed as coming at the price of nationhood. Warlaumont (2010) points out that some nations, such as France, are more protectionist in nature and this outlook shapes the very culture pervading architectures of their digital platforms. Numerous state media policies apply pressure to take on a more 'French' cultural form.

> For instance, after being allowed to maintain tariffs and quotas to protect its cultural market from other cultural products, especially American films and television, in 2003 France consumed only 60 per cent of American film products as opposed to 85 per cent in other European film markets (Riding, 2003, p. 9B). In an effort to preserve the French language the French government enacted the Toubon Law (penned in

1994 by French Cultural Minister, Jacques Toubon), mandating the use of the French language in all official and commercial publications, and imposing fines on the French media for using Americanisms or English where French equivalents exist. A few years later, another law mandated quotas requiring the majority of songs on the radio be in French (Wikipedia). Hence, these efforts were a way for the French government to put a cog in the wheel of the perceived 'steamrolling' of the English language. (Warlaumont, 2010, pp. 205–206)

While new media technologies foster new digital network spaces, they still have to comply with traditional, national and cultural policies that have dictated older communication platforms. To be exempt from these policies would mean making a case that these architectures are novel and radically divergent from existing communication platforms. Such a case is tremendously difficult in this deeply interconnected and mediated era, not just online and offline but between different old and new social networks. What is daunting, however, for instance, in the case of France, is that as cultural boundaries get tightened, command centers such as Facebook gain membership across borders. France's Facebook membership alone increased a phenomenal 518 percent over the course of 2008. Google's YouTube was the top-ranked video site in France, with 25 million people watching 2.3 billion videos online in May 2008. This was despite concerns by French government officials, who claim these particular sites have the strongest potential for American dominance and imperialism ('Instant Messaging Most Popular Online Activity in France,' 2009).

Also, we cannot escape the fact that labor circulation, however global it appears to be, is one of the most guarded territories of the nation-state. The seemingly free and voluntary activities of user labor for content production are hardly impervious to complex labor relations. This is partly framed by legal systems that are nationally based. Adam Fish (2011) expounds on the nature of labor within these transnational leisure networks. He argues that to address this critically, we need to go beyond the staid binary of viewing these practices as either celebratory of the democratic will or deeply exploitative. Instead, he suggests:

> for an inductive model of analysis that considers these two perspectives within the context of practices within the sites or systems themselves. This is all the more important when one analyzes emerging social enterprises, which attempt to fulfill a primary social, non-capitalist outcome while still maintaining a competitive position within the market. (p. 468)

This is framed as part of a system of 'creative capitalism' that "merges leisure, freedom, even fantasy with the realities of serving a competitive capitalist firm" (Fish, 2011, p. 468). Fish does underline the growing intellectual property rights challenge in the digital global network: these digital laborers are being viewed more as 'outsourced workers' as this phenomenon becomes less marginal. Take the recent case of the Nyan Cat and the lawsuit against Warner Brothers. The Nyan Cat is a feline-themed Internet meme. The video depicts an online character with a cat's face and a body resembling a horizontal breakfast bar with pink frosting. It flies across the screen, leaving a bright rainbow trail behind. Much like several YouTube phenomena that move in mysterious ways, this particular video became a sensation and went viral. Later on, Warner Brothers's game Scribblenauts used the Nyan Cat in their video game without the creator's permission. They are now facing a federal lawsuit for infringement of copyrights and trademarks. We can expect more of these incidents in the near future as boundaries of labor and leisure, and private and public become deeply blurred in this digital sphere. While Fish takes on a position of hybridity, scholars such as Terranova (2000) and Deuze (2007) point to the underlying ideology dictating these architectures, that which is more corporatized and capitalistic in nature. These spaces are seen as being far from the celebrated digital global market that promised a fair playing field for the digital laborer.

We should not forget that these social network sites, in spite of their dominant leisure and social properties, are also markets that follow economic and legal policies. Labor activity can simultaneously be play, contributing to the 'global intellectual economy.' But it is still up for contention whether one can demand an economic value for this effort. Grossman (2006) argues that we need to examine power relations between all involved actors, including the "non-monetary, social economies, and their central and increasingly constitutive role in monetary or financial economies" (Banks & Humphreys, 2008, p. 402). In other words, to determine whether labor infringements are made, we have to examine if this is a consensual and collaborative relationship. We need to recognize any form of compensation that is jointly agreed upon before we make it a matter of dispute. After all:

> the power derived from the social economies is not necessarily consonant with that derived from financial economies. When there are abrasive encounters as we have described, it is not always clear who is in control and who the winners and losers are, but it is clear that it is not as straightforward as corporate winners and user losers. Here we need a better understanding of the agents and agencies emerging through social network markets. (Banks & Humphreys, 2008, p. 413)

Last but not least, it is impossible to not address the underground global economy stimulated by digital leisure platforms. This includes the pornography industry and the impact it has on globalization flows across these spaces. Matthew Zook (2003) has examined datasets on the location of the online adult industry's content production, websites, and hosting in order to gauge the extent to which these electronic spaces interact with the geographic realm. He pays heed to the local histories, cultures, and politics of these occupied terrains:

> The roles of these actors, however, are not simply determined by a spaceless logic of cyber interaction but by histories and economies of the physical places they inhabit. In short, the 'space of flows' cannot be understood without reference to the 'space of places' to which it connects. This geography also provides a valuable counterpoint to mainstream electronic commerce and highlights the ability of socially marginal and underground interests to use the Internet to form and connect in global networks. (p. 1261)

Much in line with the global city literature, certain cities are more permissive than others in their allowances of pornography. This creates diverse spaces of regulation based on specific states and cities. For instance, Los Angeles and Amsterdam are more permissive and less litigious towards the adult industry. Their unique localization and policy influence the digital leisure space and skew it in their direction. There is also an impact felt on the nature of services offered and the means of distribution and production of adult content. Hence, we need to keep in mind that there are three geographically relevant measures to understand the impact of globalization and the physical location on digital leisure networks. These include: (1) the production of content for the industry, (2) the creation and maintenance of websites to distribute content, and (3) the hosting of websites (Zook, 2003).

In addition, much like the trajectory of clusters and the core–periphery approach as evinced in the global city literature, there is a slow but growing shift from the Western command centers to periphery cities in digital adult space. While 70 percent of the adult content distributors are based in the United States, East European actors are becoming central to the making of this industry. They are becoming key players in the business of exploitation within this global adult industry. Also, countries such as Thailand and Hungary have recently become new centers for the creation of pornography.

Hence, while the Internet has been celebrated for its connectivity across national borders, few have paid attention to the underground economy within it. Here, democratic participation takes on a new meaning, as the digital product that flows between these borders is often marked as illegal depending on the nation-state. Whether it is the fostering of terror networks or drugs or human trafficking, these traditional forms of social and mass deviance have managed to capitalize on the affordances of digital leisure networks. They use these networks to enhance their circulations, connections, and communications. Thus, Zook remarks, "the geographical manifestation of these networks shares a similar structural logic with the global financial system, albeit with drastically different priorities, goals, and relevant places" (p. 1263).

In fact, early on Castells (1998) recognized that there is a space of flows that takes on a "position of irrelevance" (p. 162) and is treated as invisible. This is true no matter how dominant its presence is. Yet, these excluded criminal realms are very much part of the global flow and contribute to the geography of Web 2.0. In essence, these deviant spheres benefit from the fragmented nature of the transnational digital topography. Such decentralization protects these realms and their participatory members against any one dominant nation-state. In some sense, globalizing of digital space creates invisibility.

As new ways of engineering of information emerge and get appropriated, they also transform boundaries between social networking sites and the Internet as we know it. In recent news (Marcus, 2013), Facebook announced its new Graph Search, which gives users the ability to search their social networks for things like photos and restaurant recommendations:

> For a long time, search has been one of the weaknesses of the site; now it might actually be useful. You can search for all the photos that you have "liked," or all the restaurants in San Francisco that have been liked by your friends who like Lady Gaga. It pores over individualized information that no other search engine has; more interestingly, it represents a different way of thinking about searching, stressing the integration of information across your posts (and those of your friends), rather than just returning a particular page that seems to fit your criteria. ('Facebook and the Future of Search': The New Yorker, January 16th 2013)

Through this lens, we might get the impression that the cityscape is becoming more a playscape. However, much innovation for remodeling the underlying infrastructures has failed to take root, giving further credence to traditional architectures of command centers. Inevitably, though, new models will threaten the relative stability of these structures, as they are still subject to digital and material movements that are political, economic, and social in nature.

Lastly, much has been written on the issue of privacy in Chapter 4. This has become a transnational concern that pervades across cultures, pushing for stringent policy measures by the state. We see divergent schools of thought emerging, including within the United States and Europe. For the most part, the United States align with a more voluntary and less state-regulatory approach, believing that the digital sphere is still in its nascent stage and its future potential would be impeded by privacy regulations. This would freeze the architectures to a high degree and have an adverse impact on innovation. From the European point of view, while innovation and economic prosperity is essential, they must have the consumer's interest at heart. After all, they are the public and these architectures should be steered in a direction that is less violating of individual rights. Over the last decade, consumers in both of these regions have demonstrated deep concern about their privacy online, triggered by the current media exposure of systematic state surveillance.

Yet, if we are to shift our attention to the emerging markets, another transnational public sphere emerges, one that appears to be less concerned about privacy. It a recent study, it was found that Saudi Internet users were at the top of the list when it came to sharing their personal lives online ('Saudis Share Almost "Everything" Online,' 2013). According to this U.S. report on computing trends, Saudi's expressed little concern about sharing most of their day-to-day lives through status updates, photos, videos, and other links. India occupies second place in this category. According to the statistics, about 60 percent of Saudis surveyed said they share "everything" or "most things" online, compared with 15 percent in the United States and 10 percent in France. The most dominant platforms used in Saudi Arabia are Snapchat, followed by Facebook and Instagram. Hence, we need to recognize that while issues such as privacy have indeed become a dominant contention within the digital leisure networks of Europe and the United States, these cannot be uncritically transferred to other transnational spheres, particularly in the periphery domains. While the West continues to exercise disproportionate power within the global landscape, there are multiple arenas in the emerging market realm where their influence is barely visible. Adopting the template of the digital global city does not necessitate the automatic transference of Western concerns and issues.

To conclude, social network sites differentiate themselves from the Internet based on user-generated content and emphasis on social and leisure orientation. However, it is important to keep in mind that social network sites continue to be embedded within the larger Internet domain and share much of the underlying architectures. In fact, one can argue that the SNS culture is pervading and influencing the shape of the Internet. To reconstitute this in metaphorical terms, the urban park distinguishes itself from the tedium of the cityscape by positioning itself as a primarily leisure and social realm. Yet it is still part of the city's fabric and subscribes to the larger operations of social norms and legalities. Hence, there are softer boundaries between the city and the park. As urban parks take on many forms reflecting contemporary leisure practices, such as shopping arcades, malls and squares within the city confines, we are alerted to the fact that the public leisure sphere is becoming more commercialized and in need of closer scrutiny. This is much like the evolution of social networking sites: corporate interests have seeped into seemingly innocuous social spaces and are, in fact, capitalizing on the disarming nature of leisure.

In thinking about the globalizing of the Internet and social networking sites, we see the persistence of the core–periphery nodes and hierarchies. These are dictated by a host of factors that are political and corporate as well as affective and culturally based. While certain command centers persist, such as Facebook, YouTube, and Twitter, we also see the rise of a host of digital leisure networks that are becoming more issue, interest and location based such as Nextdoor, BlackPlanet, BeautifulPeople, and the like. We also witness the hybridization and indigenization of dominant global leisure networks such as Facebook. In a sense, the more global these command centers get, the more nebulous their ideologies become: it is to the company's interest to have loose and fluid norms to enable the transnational scaling of their networks.

Additionally, we can argue that these social networking sites are far from disembedded from the nation-state. They seem to nurture a complex relationship where they become web representations of the nation's culture. Several new digital leisure platforms have been designed for specific regions and audiences, such as South Korean's Cyworld, Latin America's Migente, and Germany's Studivz, as well as Google's Orkut (initially aimed for the United States audience to compete with MySpace and Facebook, but it eventually took hold in Brazil). That said, the literature urges us to view these command centers not as individual nodes of power but as strategic clusters of circulating networks and capital. We need to consider how digital leisure cannot be boxed within the online world. As participants mediate between the physical and the virtual, they transform offline moments into digital memories to be consumed and played with. The digital regurgitates the past as an infinite and affective present. As projects are under way for making cities 'smart' and 'fluid' so are their leisure counterparts. Hence, we see virtual museums and malls spring up. While new information and communication technologies make possible the impressive blurring between reality and fantasy, the real and the virtual, we should not forget that many of the world's inhabitants reside in a pre-digital world. They are the invisible publics that have somehow slipped past the database that appears omnipresent. Poverty, rurality, criminality, and the perverse gain little attention within this larger discourse on the globalizing of the Internet. Going back to the metaphor, it is much like examining a city without taking heed of its vast slums—the places where half the inhabitants live, work, and play. Hence, let us use this opportunity to simultaneously enrich the conceptualizing of the city and the park, leisure and labor, and the virtual and the material by being more encompassing of the marginal and the diverse.

8 Conclusion
From Parks to Green Infrastructures

Today, many people and organizations are talking about another type of infrastructure that is critical to the continuance and growth of a community: green infrastructure.

Mark A. Benedict and Edward T. McMahon,
Green Infrastructure: Smart Conservation for the 21st Century

No single park, no matter how large and how well designed, would provide the citizens with the beneficial influences of nature. [Instead parks need to be] linked to one another and to surrounding residential neighborhoods.

Frederick Law Olmsted Jr., 1903, in
Green Infrastructure for Landscape Planning

The public park promises urban renewal. The more densely populated a city becomes, ironically, the more pressure it experiences to carve out non-instrumental public space from within. This book reveals several examples to validate this point, such as the birthing of the Massachusetts parks of the 19th century. Here, the exponential flow of immigrants into the city provoked urban planners to carve out recreational space for all. This was intended to serve as a safety valve in society. Density of human networks, it seems, require breathing room. Public space is not created for purely utilitarian purposes but caters to our deep need for experiencing leisure and pleasure. One can argue that parks are the closest that society gets to materialize its idealism. In spite of events time and again that shatter these ideals, these symbolic landscapes demonstrate an impressive resilience in the social imagination. The urban commons insist on being the common good. After all, leisure topographies fundamentally represent our humanity. The less regulated the park, the more it relies on the good in human nature to sustain it.

By now, the reader has hopefully become habituated with the practice of transcribing 'park' discussions through the lens of Web 2.0. Nobody could seriously argue that social media spaces are dominantly pragmatic. The utilitarian aspect of the digital commons sits on the sidelines while the more central need to express, connect, play, and make meaning take over. Social networks congregate online to often perform the peripheral and not the essential. Of course, embedded in the superficial fabric of play space can be vigorous digital labor, or online protest or corporate maneuverings, as we have seen throughout this book. While some scholars lean on the argument that the leisure commons dull the senses to make audiences more passive, this book does not subscribe to that belief. Granted, the romance with new media space has been tempered by decades of social practice that shows more of the bad and the ugly. The convergence of corporate giants and the state can be an intimidating prospect to say the least. Yet we see how little it takes to rekindle that romance, often through humor, play, and tactical frivolities of the people on a quotidian basis.

So we have it here in the open—while this book reveals both structure and agency of multiple digital and physical leisure networks, it hopes to have demonstrated a distinct humanist bias. Even in chapters such as Chapter 4, where these networks are positioned as architectures of fear with massive social cleavages, there is always a counterweight in human agency. Movements such as joint governance and community gardens are celebrated a little more loudly. Hackers and social activists are given heavier consideration. The marginalized are of particular focus in this book to reveal how they transform intended and often alienating public leisure space into arenas of their own making, usurping them for their own purposes.

One may ask, having drawn these parallels between the diversity of parks and Web 2.0 spaces—where do we go from here? This book requests not to move forward but to stand still. In fact, it asks the reader to slowly digest this metaphorical comparison between what is often perceived as an innocuous space—that of the public park and its more glamorous counterpart in the digital sphere. This is an unlikely comparison that does not elicit immediate inspiration, such as that of the 'frontiers' and the 'Wild Wild West,' or intense concern as with the 'electronic ghettos.' Even the technoscapes analogy can be disorienting where it emphasizes the expansiveness and borderlessness of the global flows. In contrast, the 'public park' starts out as an unremarkable metaphor and may even throw us off at the beginning. This book hopes that by now it has succeeded in demonstrating the rich and fantastical past of this radical public leisure spatialization across nations and cultures and the astounding overlap between these two spheres.

The more you delve into the park arena, the more complex their formations appear to be. It is a tribute to our social imagination on what constitutes as public leisure space. The range of ways to design, architect, sustain, and transform space relies on the politics of human action. The deeper you dig, the more natural the commonalities appear to be. The 'walled gardens' metaphor is fortuitous as it has been used in new media studies for decades. The 'fantasy parks' build on the popular critique of the Disneyfication of society to capture the commodification and commercialism of public arenas worldwide; an easy crossover. Others are not that intuitive. 'Protest parks' make sense only after the historical narrative plays out, revealing the Californian Ideology that influenced both the urban and the digital domain of the time. With the 'global parks' metaphor, a strong foundation was already in place with Mitchell's classic text on the soft city and Saskia Sassen's compelling conceptualization of global cities.

Basically, this book does not give directions to the future of new media spaces. If anything, the direction is back to the past, not of the artifact but of social spaces evoked by the artifact. In an era where scholarship on new media may be perceived as becoming obsolete with fast-paced digital innovations, this research serves as a backlash to that raw fear. For instance, does every change in Facebook's privacy settings indicate that our analysis based on past designs become inapplicable? Hardly, as the history of social space, in particular, the leisure commons, allows us to pause and pontificate on the astounding persistence of human action over centuries and across global settings. Another discipline that has been refreshing in this study is that of architecture. Architects, of all practitioners, know that designed space is rarely the executed space. In other words, lived space barely resembles intended design. Yet, architects do their best to predict human responses, mass movements, and affective behaviors and how that would influence the playing out of a spatial construct. Web 2.0 architects are not that different, and bringing these two together has been tremendously fruitful. Of course it goes without saying that the field of geography pervades and even dominates in this book, lending a backbone to multiple spatial discussions.

Lastly, we have been talking of public parks as demarcated spaces of fantasy, of protest, or of secure terrain. Yet, when speaking of these spaces, the politics of multiple institutions come to the fore. We start to see distributed networks and their interdependencies. We begin to recognize the building blocks of the park do not start and end with its spatial terrain but with the morality of the time, of pervading social values, and of the flow of people and finance. Time and again, chapter after chapter, we see how these notions parallel the media ecosystems. These vibrant architectures, both digital and urban, function as a complex social infrastructure.

As early as 1903, one of the most famous landscape architects, Frederick Law Olmsted, remarked that "a connected system of parks and parkways is manifestly far more complete and useful than a series of isolated parks" (Little, 1989, p. 77). Interestingly, in contemporary conservation literature, there is much talk of the greening of rooftops and vacant lots between high-rises. There is much vigor for converting abandoned railways and past infrastructures that stubbornly occupy urban space into parks (think of the success of the New York Chelsea High Line). Mark Benedict and Edward McMahon of the Conservation Fund describe this as *green infrastructures*, an "ecological framework for environmental, social, and economic health, in short, our natural life-support system" (2001, p. 5). It basically requires an engineering of multiple facets and the navigating of diverse terrain such as dealing with the law, the economy, and the cultural norm. Interestingly, the term 'infrastructure' connotes a very pragmatic frame of development and is commonly associated with roads, railways, electric lines, and other such networks. Yet times are changing. The utilitarian domain is getting softer around its edges and the leisure domain is getting more structured.

To conclude, in line with the spirit of this book, we can steal a page from the Green Infrastructure movement to plan and potentially address future digital leisure networks. In doing so, it is recommended that instead of viewing parks or social media spaces as parcels of terrain, we need to recognize them as networks of open space. Benedict and McMahon point out that sadly, we continue to approach the development of green infrastructures in a haphazard way.

They propose that successful land conservation in the future will have to be:

- More proactive and less reactive
- More systematic and less haphazard
- Multifunctional, not single purpose
- Large scale, not small scale
- Better integrated with other efforts to manage growth and development (2001, p. 3)

As well as:

- Systematic, not haphazard
- Holistic, not piecemeal
- Multi-jurisdictional, not single jurisdictional
- Multifunctional, not single purpose
- Multiple scales, not single scale (p. 13)

Overall, the value of green or digital leisure infrastructures lies in their interconnectedness through smart systems, driven by values that strive to keep these structures open and cared for by their consuming publics. The leisure commons, it appears, will continue to struggle to be for the common good.

References

Adams, P. (1997). Cyberspace and virtual places. *Geographical Review, 87,* 155–171.

Adams, P. (2005). *The boundless self: Communication in physical and virtual spaces.* Syracuse, NY: Syracuse University Press.

Alesina, A., Glaeser, E.L., & Sacerdote, B. (2005). *Work and leisure in the U.S. and Europe: Why So Different?* In *NBER Macroeconomics Annual 2005, 20* (pp. 1–100). Cambridge, MA: MIT Press.

Alexa Rankings (2012). Retrieved from: http://www.alexa.com/topsites.

Ali, A. (2013). Cooperative antagonism in the marketplace. *International Journal of Commerce and Management, 23*(2). Retrieved from: http://www.emeraldinsight.com/journals.htm?issn=1056- 9219&volume=23&issue=2&articleid=170 90596&show=html.

Alpizar, K., Islas-Alvarado, R., Warren, C.R., & Fiebert, M.S. (2012). Gender, sexuality and impression management on Facebook. *International Review of Social Sciences and Humanities, 4*(1), 121–125.

Alrumaihi, M. (2013). Arab Spring: Battle for free speech. *Gulf News.* April 6. Retrieved from: http://m.gulfnews.com/opinion/arab-spring-battle-for-free-speech-1.1167471.

Amin, A. (2005). Local community on trial. *Economy and Society, 34*(4), 612–633.

Amin, A. (2008). Collective culture and urban public space. *City, 12*(1), 5–24.

Anderson, S.E. (2008). *Digital enclosure and the communication commons revival.* (Doctoral dissertation). Retrieved from: http://summit.sfu.ca/item/8744.

Andrejevic, M. (2011). The work that affective economics does. *Cultural Studies, 25*(4–5), 604–620.

Appadurai, A. (1996). *Modernity at large: Cultural dimensions of globalization.* Minneapolis, MN: University of Minnesota Press.

Appadurai, A. (2013). *The Future as cultural fact: Essays on the global condition.* New York, NY: Verso Books.

Arora, P. (2010). *Dot com mantra: Social computing in the Central Himalayas.* Burlington, VT: Ashgate Publishing.

Arora, P. (2011). Online social sites as virtual parks: An investigation into leisure online and offline. *Information Society, 27*(2), 113–120.

Arora, P. (2012a). Leisure divide: Can the third-world come out to play? *Information Development, 28*(2), 93–101.

Arora, P. (2012b). Typology of Web 2.0 spheres: Understanding the cultural dimensions of social media spaces. *Current Sociology, 60*(5), 599–618.

Arora, P., & Rangaswamy, N. (2013). Digital leisure for development: Rethinking new media practices from the global south. *Media Culture & Society, 35*(7), 898–905.

Arora, P. (2014). Usurping public leisure space for protest: Social activism in the digital and material commons. *Space and Culture*, 1–26: doi: 10.1177/1206331213517609

Atkinson, R., & Blandy, S. (2005). The new enclavism and the rise of gated communities. *Housing Studies, 20*(2), 177–186.

Aufderheide, P. (2002). Competition and commons: The public interest in and after the AOL–Time Warner merger. *Journal of Broadcasting & Electronic Media, 46*(4), 515–531.

Austin, G. (2014). *Green infrastructure for landscape planning: Integrating human and natural systems.* New York: NY: Routledge.

Ayres, I. (2007). *Super crunchers: How anything can be predicted.* London, UK: John Murray.

Bakhtin, M.M. (1984). *Problems of Dostoevsky's poetics* [ed. and trans. C. Emerson]. Minneapolis, MN: University of Minnesota Press.

Bakir, V. (2010). *Sousveillance, media and strategic political communication: Iraq, USA, UK.* London, UK: Continuum International Publishing Group.

Balkin, J.M. (2004). Virtual liberty: Freedom to design and freedom to play in virtual worlds. *Virginia Law Review, 90*, 2043–2098.

Banks, J., & Humphreys, S. (2008). The labour of user co-creators emergent social network markets? *Convergence: The International Journal of Research into New Media Technologies, 14*(4), 401–418.

Barbrook, R., & Cameron, A. (1996). The Californian ideology. *Science as Culture, 6*(1), 44–72.

Barlow, J.P. (1996). *Declaration of independence for cyberspace.* Retrieved from: http://wac.colostate.edu/rhetnet/barlow/barlow_declaration.html.

Barnes, P. (2006). *Capitalism 3.0: A guide to reclaiming the commons.* San Francisco, CA: Berrett-Koehler.

Barnes, S.B. (2006). A privacy paradox: Social networking in the United States. *First Monday, 11*(9). Retrieved from: http://firstmonday.org/htbin/cgiwrap/bin/ojs/index.php/fm/arti- %20cle/view/1394/1312.

Barnett, G.A., & Park, H.W. (2005). The structure of international Internet hyperlinks and bilateral bandwidth. *Annales des telecommunications, 60*(9–10), 1110–1127.

Bartle, R.A. (2010). From MUDs to MMORPGs: The history of virtual worlds. In J. Hunsinger, L. Klastrup and M. Allen (Eds.), *International handbook of Internet research* (pp. 23–39). Amsterdam, NL: Springer.

Baudelaire, C. (1964). *The painter of modern life and other essays* [trans. J. Mayne]. London, UK: Phaidon.

Baym, N.K. (2009). A call for grounding in the face of blurred boundaries. *Journal of Computer-Mediated Communication, 14*(3), 720–723.

Benedict, A., & McMahon, E.T. (2001). *Green infrastructure: Smart conservation for the 21st century.* Washington, DC: The Conservation Fund: Sprawl Watch Clearing House Monograph Series.

Benedikt, M. (Ed.) (1991). *Cyberspace: First steps.* Cambridge, MA: MIT Press.

Benjamin, R.I., & Wigand, R.T. (1995). Electronic markets and virtual value chains on the information superhighway. *Sloan Management Review (Winter).* Retrieved from: http://sloanreview.mit.edu/article/electronic-markets-and-virtual-value-chains-on-the- information-superhighway/.

Benjamin, W. (1983). *Charles Baudelaire: A lyric poet in the era of high capitalism.* London, UK: NLB.

Benkler, Y. (2006). *The wealth of networks: How social production transforms markets and freedom.* New Haven, CT: Yale University Press.

Bennett, W.L. (2012). The personalization of politics: Political identity, social media, and changing patterns of participation. *ANNALS of the American Academy of Political and Social Science, 644,* 20–39.

Boellstorff, T. (2010). *Coming of age in second life: An anthropologist explores the virtually human.* Princeton, NJ: Princeton University Press.

Bollier, D. (2002). Reclaiming the commons. *Boston Review.* Retrieved from: http://bostonreview.net/BR27.3/bollier.html.

Boyd, D.M. (2007). Why youth (heart) social network sites: The role of networked publics in teenage social life. In D. Buckingham (Ed.), *MacArthur Foundation series on digital learning—youth, identity, and digital media volume* (pp. 119–142). Cambridge, MA: MIT Press.

Boyd, D.M. (2011).White flight in networked publics? How race and class shaped American teen engagement with MySpace and Facebook. In L. Nakamura and P.A. Chow-White (Eds.), *Race after the Internet* (pp. 203–222). London, UK: Routledge.

Boyd, D.M. (2013). *How would you define work in a networked world?* May 20. Retrieved from: http://www.zephoria.org/thoughts/.

Boyd, D.M., & Ellison, N.B. (2007). Social network sites: Definition, history and scholarship. *Journal of Computer-Mediated Communication, 13*(1), 210–230.

Boyd, D.M., & Marwick, A. (2011, September). Social privacy in networked publics: Teens' attitudes, practices, and strategies. Paper presented at the Oxford Internet Institute. Retrieved from: http://www.danah.org/papers/2011/Social-PrivacyPLSC-Draft.pdf.

Brenner, N. (1998). Global cities, glocal states: global city formation and state territorial restructuring in contemporary Europe. *Review of International Political Economy, 5*(1), 1–37.

Brenner, N., & Keil, R. (Eds.) (2006). *The global cities reader.* London, UK: Routledge.

Broudehoux, A. (2007). Spectacular Beijing: The conspicuous construction of an Olympic metropolis. *Journal of Urban Affairs, 29*(4), 383–399.

Bruner, M.L. (2005). Carnivalesque protest and the humorless state. *Text and Performance Quarterly, 25*(2), 136–155.

Buckingham, D., & Willett, R. (2006). *Digital generations: Children, young people, and new media.* Hillsdale, NJ: Lawrence Erlbaum Associates, Publishers.

Burbules, N.C. (2002). The web as a rhetorical place. In I. Snyder (Ed.), *Silicon literacies: Communication, innovation and education in the electronic age* (pp. 75–84). London, UK: Routledge.

Caldeira, T.P.R. (2000). *City of walls: Crime, segregation and citizenship in São Paulo.* Berkeley, CA: University of California Press.

Cameron, D., & Stein, J.G. (Eds.) (2002). *Street Protests and Fantasy Parks: Globalization, Culture, and the State.* Vancouver, Canada: UBC Press.

Campbell, J.E. (2004). *Getting it on online: Cyberspace, gay male sexuality, and embodied identity.* New York, NY: Harrington Park Press.

Campbell, S. (1987). *Cottesbrooke: An English Kitchen Garden.* Salem, MA: Salem House.

Canter, D., Rivers, R., & Storrs, G. (1985). Characterizing user navigation through complex data structures. *Behaviour & Information Technology, 4,* 93–102.

Carlton, L. (2011). Video games become a hot trend in real life theme park attractions. *FoxNews.com.* September 19. Retrieved from: http://www.foxnews.com/travel/2011/09/19/video-games-theme-parks/.

Carpenter, M.A., & Sanders, W.G. (2007). *Strategic management: A dynamic perspective.* Upper Saddle River, NJ: Pearson/Prentice Hall.

Carroll, W.K. (2007). Global cities in the global corporate network. *Environment and Planning A, 39*(10), 2297–2323.

Carter, M. (2009, April 16). TV and social media: Fighting for dominance. *New Media Age*, 16–17.

Castells, M. (1996). *The rise of the network society. The Information Age: Economy, Society, and Culture Vol. I.* Cambridge, MA: Blackwell.

Castells, M. (1998). *End of millennium, The Information Age: Economy, Society, and Culture Vol. III.* Cambridge, MA: Blackwell.

Castells, M., & Hall, P. (1994). *Technopoles of the world: The making of 21st century industrial complexes.* London, UK: Oxford University Press.

Castronova, E. (2001). Virtual worlds: A first-hand account of market and society on the cyberian frontier. *The Gruter Institute Working Papers on Law, Economics, and Evolutionary Biology*, 2(1). Retrieved from: http://www.bepress.com/giwp/default/vol2/iss1/art1/.

Cavallo, D. (1981). *Muscles and morals: Organized playgrounds and urban reform, 1880–1920.* Philadelphia, PA: University of Pennsylvania Press.

Certeau, M. de. (1984). *The practice of everyday life.* Berkeley, CA: University of California Press.

Chance, H. (2012). Mobilising the modern industrial landscape for sports and leisure in the early twentieth century. *International Journal of the History of Sport*, 29(11), 1600–1625.

Chang, J. (2006). Behind the glass curtain: Google's new headquarters balances its utopian desire for transparency with its very real need for privacy. *Metropolis Magazine*. June 19. Retrieved from: http://www.metropolismag.com/story/20060619/behind-the-glass-curtain.

Chengpeng , Li. (2013). *The Whole World Knows.* Xinxing (New Star Press), Beijing.

Cho, S., & Huh, J. (2010). Content analysis of corporate blogs as a relationship management tool. *Corporate Communications: An International Journal*, 15(1), 30–48.

Chudacoff, H.P. (2007). *Children at play: An American history.* New York, NY: New York University Press.

Clavé, S.A. (2007). *The global theme park industry.* Cambridge, MA: CABI.

Cohen, E.J. (2007). Cyberspace as/and space. *Columbia Law Review*, 107, 210–256.

Coopman, M.T. (2011). Networks of dissent: Emergent forms in media based collective action. *Critical Studies in Media Communication*, 28(2), 153–172.

Curry, M.R. (2005). Toward a geography of a world without maps: Lessons from Ptolemy and postal codes. *Annals of the Association of American Geographers*, 95(3), 680–691.

Daniels, J. (2012). Race and racism in Internet studies: A review and critique. *New Media & Society*, 15(5), 695–719.

Dery, M. (1993). *Culture jamming: Hacking, slashing, and sniping in the empire of signs.* Open Magazine Pamphlet Series: NJ. 'Shovelware.' Retrieved from http://markdery.com/?page_id=154

D'Arcus, B. (2006). *Boundaries of dissent: Protest and state power in the media age.* New York, NY: Routledge.

Daskalaki, M., Starab, A., & Imasa, M. (2008). The 'parkour organisation': Inhabitation of corporate spaces. *Culture and Organization*, 14(1), 49–64.

Davis, I. (1953). Urban farming. A study of the agriculture of the city of Birmingham. *Geography*, 38(4), 296–303.

Davis, D.S., Kraus, R., Naughton, B., & Perry, E.J. (1995). *Urban spaces in contemporary China: The potential for autonomy and community in post-Mao China.* Cambridge, MA: Cambridge University Press.

della Porta, D. (2005). Multiple belongings, tolerant identities, and the construction of 'another politics': Between the European social forum and the local social fora. In D. della Porta and S. Tarrow (Eds.), *Transnational protest and global activism* (pp. 175–202). Lanham, MD: Rowman and Littlefield.

Derudder, B., Taylor, P., Ni, P., De Vos, A., Hoyler, M., Hanssens, H., Bassens, D., Huang, J., Witlox, F., Shen, W., & Yang, X. (2010). Pathways of change: Shifting connectivities in the world city network, 2000–08. *Urban Studies, 47*(9), 1861–1877.

Dery, M. (2010 [1993]). *Culture jamming: Hacking, slashing, and sniping in the empire of signs.* NJ: Open Magazine Pamphlet Series, Open Magazine.

Deuze, M. (2007). *Media work.* Cambridge, UK: Polity.

Devree, C. (1957). Private life of a park. *New York Times*, quoted in 'Gramercy Park Historic District' at the NYC Landmarks Preservation Commission. December 8. Retrieved from: http://www.nyc.gov/html/lpc/downloads/pdf/reports/GRAMERCY_PARK_HISTORIC_DISTRICT.pdf.

DiPrete, T.A., Gelman, A., McCormick, T., Teitler, J., & Zheng, T. (2011). Segregation in social networks based on acquaintanceship and trust. *American Journal of Sociology, 116*(4), 1234–1283.

Disney, W. (n.d). Sharing the Magic of Disney. Retrieved from: http://disneyheaven.com/WaltDisneyWorld.htm

Dodge, M., & Kitchin, R. (2000). *Mapping cyberspace.* London, UK: Routledge.

Donath, J. (2007). Signals in social supernets. *Journal of Computer-Mediated Communication, 13*(1), 231–251.

Dron, J. (2006). Social software and the emergence of control. In R. Koper, P. Kommers, P.A. Kirschner, D. Sampson and W. Didderen (Eds.), *Proceedings of the Sixth International Conference on Advanced Learning Technologies* (pp. 904–908). New York, NY: IEEE.

Du Gay, P. (1996). *Consumption and identity at work.* London, UK: Sage.

Duncombe, S. (2012). Imagining no-place. In H. Jenkins and S. Shresthova (Eds.), "Transformative Works and Fan Activism", special issue, *Transformative Works and Cultures 10.* Retrieved from: http://journal.transformativeworks.org/index.php/twc/article/view/350/266.

Dupont, V.D. (2011). The dream of Delhi as a global city. *International Journal of Urban and Regional Research, 35*(3), 533–554.

Edwards, J. (2004). *Blogging for Dollars: Once the domain of the disgruntled and demented, Web logs are being embraced by business executives.* CFO Magazine. Retrieved from: http://www.cfo.com/printable/article.cfm/3238412?f=options

Ellin, N. (Ed.) (1997). *Architecture of fear.* New York, NY: Princeton Architectural Press.

Elwood, S. (2002). Neighborhood revitalization through 'collaboration': Assessing the implications of neoliberal urban policy at the grassroots. *GeoJournal, 58*(2–3), 121–130.

European Commission. (2010). *Communication from the Commission to the European Parliament, the Council, the European Economic and Social Committee and the Committee of the Regions: a Digital Agenda for Europe.* Retrieved from: http://eur- lex.europa.eu/LexUriServ/LexUriServ.do?uri=com:2010:0245 :fin:en:pdf.

Everett, A. (Ed.) (2008). *Learning race and ethnicity: Youth and digital media.* Cambridge, MA: MIT Press.

Falzon, M. (2004). Paragons of lifestyle: Gated communities and the politics of space in Bombay. *City & Society, 16*(2), 145–167.

Ferguson, S. (1999). A Brief History of Grassroots Greening on the Lower East Side. In P.L. Wilson and B. Weinberg (Eds.), *Avant Gardening: Ecological Struggle in the City and the World* (pp. 60–79). New York, NY: Autonomedia.

Fischer, C.S. (1994). Changes in leisure activities, 1890–1940. *Journal of Social History, 27*(3), 453–475.

Fish, A. (2011). Governance of labor in digital video networks. In *Proceedings of the 2011 iConference* (pp. 466–471). New York, NY: ACM

Florida, R. (2002). *The rise of the creative class and how its transforming work, leisure, community & everyday life.* New York, NY: Basic Books.

Forced into extinction. (2013). *The Economist.* January 21. Retrieved from: http://www.economist.com/blogs/pomegranate/2013/01/saudi-arabia.

Fraser, N., & Nash, K. (2013). *Transnationalizing the public sphere.* Cambridge, UK: Polity Press.

Freudenheim, M. (2012). Digitizing health records, before it was cool. *New York Times.* January 14. Retrieved from: http://www.nytimes.com/2012/01/15/business/epic-systems-digitizing-health-records-before-it-was-cool.html?pagewanted=all.

Fried, J. (2010). Why the office is the worst place to work. *CNN Opinion.* December 5. Retrieved from: http://articles.cnn.com/2010–12–05/opinion/fried.office.work_1_office-equipment-work-head-hits?_s=PM:OPINION.

Friedman, T.L. (2006). *The world is flat [updated and expanded]: A brief history of the twenty- first century.* New York, NY: Farrar, Straus & Giroux.

Frost, R. (1934). *Mending wall.* Baltimore, Maryland: Enoch Pratt Free Library.

Forsyth, A., & Crewe, K. (2010). Suburban technopoles as places: The international campus- garden-suburb style. *Urban Design International, 15*(3), 165–182.

Foucault, M. (1967). *Of other spaces: Utopias and heterotopias.* Retrieved from: http://web.mit.edu/allanmc/www/foucault1.pdf.

Fuchs, C. (2006). The self-organization of cyberprotest. In K. Morgan, C. Brebbia, A.Carlos and J. Michael Spector (Eds.), *The Internet society II: Advances in education, commerce & governance* (pp. 275–295). Boston, MA: WIT Press.

Gajjala, R., Rybas, N., & Altman, M. (2008). Racing and queering the interface producing global/local cyberselves. *Qualitative Inquiry, 14*(7), 1110–1133.

Galician, M-L. (Ed.) (2013). *Handbook of product placement in the mass media: New strategies in marketing theory, practice, trends, and ethics.* New York, NY: Routledge.

Ganesh, I.M. (2010). Mobile love videos make me feel healthy: Rethinking ICTs for development. *IDS Working Paper 352.* Brighton, UK: IDS.

Garnham, N. (2000). *Emancipation, the media, and modernity. Arguments about the media and social theory.* Oxford, UK: Oxford University Press.

Gaudiosi, J. (2012). Rovio execs explain what angry birds toons channel opens up to its 1.7 billion gamers. *Forbes.com.* March 11. Retrieved from: http://www.forbes.com/sites/johngaudiosi/2013/03/11/rovio-execs-explain-what-angry-birds-toons-channel-opens-up-to-its-1-7-billion-gamers/.

Gavison, R. (1992). Feminism and the private–public distinction. *Stanford Law Review, 45,* 1–45.

Gely, R., & Bierman, L. (2007). Social isolation and American workers: Employee blogging and legal reform. *Harvard Journal of Law and Technology, 20*(2), 288–326.

Gentzkow, M., & Shapiro, J.M. (2006). What drives media slant? Evidence from U.S. daily newspapers. *NBER Working Paper No. 12707.* Cambridge, MA: National Bureau of Economic Research.

Gentzkow, M., & Shapiro, J. M. (2011). Ideological segregation online and offline. *Quarterly Journal of Economics, 126*(4), 1799–1839.

Gershuny, J. (2005). Busyness as the badge of honor for the new superordinate working class. *Social Research: An International Quarterly, 72*(2), 287–314.

Giese, K. (2004). Speaker's corner or virtual panopticon: Discursive construction of Chinese identities online. In F. Mengin (Ed.), *Cyber China: Reshaping national identities in the age of information* (pp. 19–36). New York, NY: Palgrave.

Gitlin, T. (2002/2007). *Media unlimited: How the torrent of images and sounds overwhelms our lives*. New York, NY: Henry Holt and Company.

Gladwell, M. (2010). Small change: Why the revolution will not be tweeted. *New Yorker*. October 4. Retrieved from: http://www.newyorker.com/reporting/2010/10/04/101004fa_fact_gladwell#ixzz1gyZDFL 81.

Goldstein, H.A., & Luger, M.I. (1990). Science/technology parks and regional development theory. *Economic Development Quarterly, 4*(1), 64–78.

Gough, P. (2000). From heroes' groves to parks of peace: Landscapes of remembrance, protest and peace. *Landscape Research, 25*(2), 213–228.

Graham, L.L. (2014). *Virgin terrain: Cybercultured bodies, technology-mediated education, and online production in the digital age.* (Dissertation thesis. Columbia University, New York City.)

Graham, S., & Marvin, S. (2001). *Splintering urbanism, networked infrastructures, technological mobilities and the urban condition.* London, UK: Routledge.

Graham, S., & Marvin, S. (2005). *Telecommunications and the city: Electronic spaces, urban places.* New York, NY: Routledge.

Graham, M., Stephens, M., & Hale, S. (2013). Mapping the geoweb: A geography of Twitter. *Environment and Planning A, 45*(1), 100–102.

Grant, J., & Mittelsteadt, L. (2004). Types of gated communities. *Environment and Planning B: Planning and Design, 31*, 913–930.

Grasmuck, S., Martin, J., & Zhao, S. (2009). Ethno-racial identity displays on Facebook. *Journal of Computer-Mediated Communication, 15*(1), 158–188.

Greenleaf, G. (1998). An endnote on regulating cyberspace: Architecture vs. law? *UNSW Law Journal, 21*, 593–623.

Gross, R., & Acquisti, A. (2005). Information revelation and privacy in online social networks. In: *Proceedings of the 2005 ACM workshop on Privacy in the electronic society* (pp. 71- 80). New York, NY: ACM.

Grossman, L. (2006). Time's Person of the Year: You. *Time.com.* December 13. Retrieved from: http://content.time.com/time/magazine/article/0,9171,1570810,00.html.

Guerrier, Y., & Adib, A. (2003). Work at leisure and leisure at work: A study of the emotional labour of tour reps. *Human Relations, 56*(11), 1399–1417.

Guillen, M.F. (1997). Scientific management's lost aesthetic: architecture, organization, and the Taylorized beauty of the mechanical. *Administrative Science Quarterly, 42*(4), 682–715.

Gunkel, A.H., & Gunkel, D. (1997). Virtual geographies: The new worlds of cyberspace. *Critical Studies in Mass Communication, 14*, 123–137.

Gunkel, A.H., & Gunkel, D. (2009). Terra nova 2.0—The new world of MMOR-PGs. *Critical Studies in Media Communication, 26*(2), 104–127.

Gupta, P., Kim, M., & Levine, D. (2013). Apple and Samsung, frenemies for life. *Reuters.* February 11. Retrieved from: www.reuters.com/article/2013/02/11/apple-samsung- idUSL1N0BA0AL20130211.

Habermas, J. (1962/1989). *The structural transformation of the public sphere: An inquiry into a category of a bourgeois society* [trans. T. Burger and F. Lawrence]. Cambridge, MA: MIT Press.

Hardin, G. (1968). The tragedy of the commons. *Science, 162*(3859), 1243–1248.

Hardt, M., & Negri, A. (2001). *Empire.* Cambridge, MA: Harvard University Press.

Hardt, M., & Negri, A. (2004). *Multitude: War and democracy in the age of empire.* London, UK: Penguin Books.

Harrison, R. (2004). *Congress, progressive reform, and the new American state.* Cambridge, MA: Cambridge University Press.

Hassid, J. (2012). Safety valve or pressure cooker? Blogs in Chinese political life. *Journal of Communication, 62*, 212–230.

Held, D., & McGrew, A. (Eds.) (2000). *The global transformations reader*. Cambridge, UK: Polity Press.

Hench, J., & Van Pelt, P. (2003). *Designing Disney: Imagineering and the art of the show*. New York, NY: Disney Editions.

Henderson, K.A. (1996). One size doesn't fit all: The meanings of women's leisure. *Journal of Leisure Research, 28*(3), 139–154.

Henley, N. (1977). *Body politics: Power, sex and non-verbal communication*. Englewood Cliffs, NJ: Prentice Hall.

Hermann, C. (2006). Laboring in the network. *Capitalism Nature Socialism, 17*(1), 65–76.

Hess, D. (1995). *Science and technology in a multicultural world: The cultural politics of facts and artifacts*. New York, NY: Columbia University Press.

Higgs, D. (Ed.) (1999). *Queer sites: Gay urban histories since 1600*. London, UK: Routledge.

Hinshaw, D. (2011). It's thrilling: The rise of Africa's amusement parks: What Africa's booming middle class really wants is a roller coaster. *Global Post*. July 14. Retrieved from: http://www.globalpost.com/dispatch/news/regions/nigeria/110713/africa-amusement-parks-dakar-african-middle-class.

Hjorth, L. (2011). *Games and gaming: An introduction to new media*. Oxford, UK: Berg.

Hook, D., & Vrdoljak, M. (2002). Gated communities, heterotopia and a 'rights' of privilege: A 'heterotopology' of the South African security-park. *Geoforum, 33*(2), 195–219.

Hospers, G.J. (2003). Creative cities in Europe. *Intereconomics, 38*(5), 260–269.

Howe, J. (2006). The rise of crowdsourcing. *Wired Magazine*. Retrieved from: http://www.wired.com/wired/archive/14.06/crowds_pr.html.

Hudson, S., & Hudson, D. (2006). Branded entertainment: A new advertising technique or product placement in disguise? *Journal of Marketing Management, 22*(5–6), 489–504.

Hunt, J.D., & Willis, P. (Eds.) (2000 [1975]). *The genius of the place: The English landscape garden, 1620–1820*. Cambridge, MA: MIT Press.

Hunter, D. (2003). Cyberspace as place and the tragedy of the digital anticommons. *California Law Review, 91*(2), 439–519.

Hunter, L. (1985, January). Public image. *Whole Earth Review*, 32–37.

Hutchison, R. (1988). A critique of race, ethnicity, and social class in recent leisure-recreation research. *Journal of Leisure Research, 20*(1), 10–30.

India's demographic challenge: Wasting time. (2013). *Economist*. May 11. Retrieved from: http://www.economist.com/news/briefing/21577373–india-will-soon-have-fifth-worlds-working-age-population-it-urgently-needs-provide.

Instant messaging most popular online activity in France. (2009). *ComScore*. April 6. Retrieved from: http://www.comscore.com/Insights/Press_Releases/2009/4/Instant_Messaging_Most_Pop ular_Online_Activity_in_France.

International Association of Amusement Parks and Attractions (IAAPA). (2013). Retrieved from: http://www.iaapa.org/resources/by-park-type/amusement-parks-and- attractions/industry-statistics.

Internet World Statistics. (2012). Retrieved from: http://www.internetworldstats.com/stats.htm

Invisible sieve: Hidden, specially for you. (2011). *The Economist*. June 30. Retrieved from: http://www.economist.com/node/18894910?story_id=18894910&fsrc=rss.

Ishida, T. (2000). Understanding digital cities. *Digital Cities, 1765*, 7–17.

Jacob, I., & Walsh, N. (2004). *Architecture of the World Wide Web, Volume One*. W3C Recommendation. Retrieved from: http://www.w3.org/TR/webarch/.

Jacobs, J. (1961). *The life and death of great American cities.* New York, NY: Random House.

Jamieson, K.H., & Cappella, J.N. (2008). *Echo chamber: Rush Limbaugh and the conservative media establishment.* New York, NY: Oxford University Press.

Jenkins, H. (2006). *Convergence culture: Where old and new media collide.* New York, NY: NYU Press.

Jiang, M., & Xu, H. (2009). Exploring online structures on Chinese government portals: Citizen political participation and government legitimation. *Social Science Computer Review, 27*(2), 174–195.

Jiao, W. (2007). E'gao: Popular art criticism or just plain evil? *China Daily.* January 22. Retrieved from: http://www.chinadaily.com.cn/cndy/2007–01/22/content_788540.htm.

Johnson, M. (2010). Metaphor and cognition. *Handbook of Phenomenology and Cognitive Science, 2,* 401–414.

Jones, L.M., Mitchell, K.J., & Finkelhor, D. (2012). Trends in youth internet victimization: Findings from three youth internet safety surveys 2000–2010. *Journal of Adolescent Health, 50*(2), 179–186.

Kafai, Y.B., Heeter, C., Denner, J., & Sun, J.Y. (2008). *Beyond Barbie and Mortal Kombat: New perspectives on gender and computer games.* Cambridge, MA: MIT Press.

Kasson, J.F. (1978). *Amusing the million: Coney island at the turn of the century.* New York, NY: Hill and Wang.

Kaupins, G., & Park, S. (2011). Legal and ethical implications of corporate social networks. *Employee Responsibility and Rights Journal, 23,* 83–99.

Kendall, L. (2002). *Hanging out in the virtual pub: Masculinities and relationships online.* Berkeley, CA: University of California Press.

Kent, S. (2010). *The ultimate history of video games: From Pong to Pokémon and beyond . . . the story behind the craze that touched our lives and changed the world.* New York, NY: Random House.

Kihlgren, A. (2003). Promotion of innovation activity in Russia through the creation of science parks: The case of St. Petersburg (1992–1998). *Technovation, 23*(1), 65–76.

Kjerulf, A. (2007). *Happy hour is 9 to 5: Learn how to love your job, create a great business and kick butt at work.* Alexander Kjerulf: http://positivesharing.com/happyhouris9to5/.

Koehler, M.J., Mishra, P., Bouck, E.C., DeSchyver, M., Kereluik, K, Seob Shin, T., & Wolf, L.G. (2011). Deep-play: Developing TPACK for 21st century teachers. *International Journal of Learning Technology, 6*(2), 146–163.

Koh, J., & Beck, A. (2006). Parks, people and city. *Topos, 55,* 14–20.

Komninos, N., Pallot, M., & Schaffers, H. (2012). Special Issue on Smart Cities and the Future Internet in Europe. *Journal of the Knowledge Economy, 4*(2), 119–134.

König, R. (2013). Wikipedia: Between lay participation and elite knowledge representation. *Information Communication & Society, 16*(2), 160–177.

Kramer, R.M., & Cook, K.S. (Eds.) (2004). *Trust and distrust in organizations: Dilemmas and approaches.* New York, NY: Russell Sage Foundation.

Krogstie, J. (2012). Modeling of digital ecosystems: Challenges and opportunities. *Collaborative Networks in the Internet of Services, 380,* 137–145.

Lakoff, G., & Johnson, M. (1980). *Metaphors we live by.* Chicago, IL: University of Chicago Press.

Lakshmi, R. (2013). On New Delhi street, protest simmers nonstop. *Washington Post.* March 30. Retrieved from: http://articles.washingtonpost.com/2013–03–30/world/38140961_1_india-gate-jantar-mantar-anger.

Latour, B. (1999). On recalling ANT. In: J. Law and J. Hassard (Eds.), *Actor network theory and after* (pp. 15–25). Oxford, UK: Blackwell.

Laurier, E. (2008). Drinking up endings: Conversational resources of the cafe. *Language & Communication, 28*(2), 165–181.

Lawrence, D. (1982). Parades, politics, and competing urban images: Doo dah and roses. *Urban Anthropology, 11*(2), 155–176.

Leach, W.R. (1993). *Land of desire: Merchants, power, and the rise of a new American culture.* New York, NY: Pantheon Books.

Leadbeater, C. (2007). *We think: Why mass creativity is the next big thing.* Retrieved from: http://www.wethinkthebook.net/home.aspx.

Leader-Chivee, L., & Cowan, E. (2008). Networking the way to success: Online social networks for workplace and competitive advantage. *People Strategy, 31*(4), 40–46.

Leighley, J.E. (Ed.) (2010). *The Oxford handbook of American elections and political behavior.* Oxford, UK: Oxford University Press.

Lefebvre, H. (1991). *The production of space* [trans. D. Nicholson-Smith]. Oxford, UK: Basil Blackwell.

Legge, K. (2005). *The rhetorics and realities of HRM, Anniversary Edition.* London, UK: Palgrave.

Lemley, M.A. (2003). Place and cyberspace. *California Law Review, 91*, 521–542.

Lessig, L. (1999). *Code and other laws of cyberspace.* New York, NY: Basic Books.

Levine, R. (2005). Geography of busyness. *Social Research, An International Quarterly, 72*(2), 355–370.

Leyshon, A. (2001). Time-space (and digital) compression: Software formats, musical networks, and the reorganisation of the music industry. *Environment and Planning A, 33*(1), 49–78.

Lindtner, S., & Szablewicz, M. (2010). China's many Internets: Participation and digital game play across a changing technology landscape. Paper presented at the China Internet Research Conference, Peking, China. Retrieved from: http://feiyaowan.files.wordpress.com/2011/09/lindtner_szablewicz-circ_final.pdf.

Little, C.E. (1989). *Greenways for America.* Baltimore, MD: Johns Hopkins University Press.

Livingstone, S. M. (Ed.). (2005). *Audiences And Publics: When Cultural Engagement Matters For The Public Sphere: Changing Media, Changing Europe* (Vol. 2). Bristol, UK: Intellect Books.

Lopez, J. (2003). *Society and its metaphors: Language, social theory and social structure.* London, UK: Continuum.

Lorimer, H. (2005). Cultural geography: The busyness of being 'more-than-representational.' *Progress in Human Geography, 29*(1), 83–94.

Loukaitou-Sideris, A., & Banerjee, T. (1993).The negotiated plaza: Design and development of corporate open space in downtown Los Angeles and San Francisco. *Journal of Planning Education and Research, 13*(1), 1–12.

Loures, L., Santos, R., & Panagopoulos, T. (2007). Urban parks and sustainable city planning: The case of Portimão, Portugal. *Wseas Transactions on Environment and Development, 10*(3), 171–180.

Low, S. (2003). *Behind the gates; Life, security, and the pursuit of happiness in fortress America.* New York, NY: Routledge.

Luke, R. (2006). The phoneur: Mobile commerce and the digital pedagogies of the wireless web. In: P. Trifonas (Ed.), *Communities of difference: Culture, language, technology* (pp. 185–204). London, UK: Palgrave Macmillan.

Lundby, K. (2011). Patterns of belonging in online/offline interfaces of religion. *Information, Communication & Society, 14*(8), 1219–1235.

MacKinnon, R. (2008). Flatter world and thicker walls? Blogs, censorship and civic discourse in China. *Public Choice, 134*(1), 31–46.

Malone, K. (2007). The bubble-wrap generation: Children growing up in walled gardens. *Environmental Education Research, 13*(4), 513–527.

Marcella Pattyn: Obituary (2013). *The Economist.* April 27. Retrieved from: http://www.economist.com/news/obituary/21576632–marcella-pattyn-worlds- last- beguine-died-april-14th-aged-92–marcella-pattyn.

Marcus, G. (2013). Facebook and the future of search. *New Yorker.* January 16. Retrieved from: http://www.newyorker.com/online/blogs/newsdesk/2013/01/ facebooks-new-science-of- search.html.

Marcuse, P., & van Kempen, R. (Eds.) (2000). *Globalizing cities: A New Spatial Order.* London, UK and Cambridge, MA: Blackwell Publishers.

Marshall, S.J., Jones, D.A., Ainsworth, B.E., Reis, J.P., Levy, S.S., & Macera, C.A. (2007). Race/ethnicity, social class, and leisure-time physical inactivity. *Medicine and Science in Sports and Exercise, 39*(1), 44–51.

Martinez-Fernandez, C., Audirac, I., Fol, S., & CunninghamSabot, E. (2012). Shrinking cities: Urban challenges of globalization. *International Journal of Urban and Regional Research, 36*(2), 213–225.

Massey, D., Quintas, P., & Wield, D. (1992). *High-tech fantasies: Science parks in society, science and space.* London, UK: Routledge.

McChesney, R.W. (2013). *Digital disconnect: How capitalism is turning the Internet against democracy.* New York, NY: New Press.

McCorkindale, T. (2010). Can you see the writing on my wall? A content analysis of the Fortune 50's Facebook social networking sites. *Public Relations Journal, 4*(3), 1–14.

McGrath, L.C. (2010). Adoption of social media by corporations: A new era. *Business and Economic Review, 13*, 14–19.

McKay, G. (2011). *Radical gardening: Politics, idealism & rebellion in the garden.* London, UK: Frances Lincoln.

McLuhan, M. (1962). *The Gutenberg Galaxy: the making of typographic man.* Toronto, Canada: University of Toronto Press.

Mejias, U. (2010). The Twitter revolution must die. *International Journal of Learning and Media, 2*(4), 3–5.

Mitchell, D. (1995). The end of public space? People's park, definitions of the public, and democracy. *Annals of the Association of American Geographers, 85*(1), 108–133.

Mitchell, W.J. (1996). *City of bits: Space, place, and the infobahn.* Cambridge, MA: MIT Press.

Mitrasinovic, M. (2006). *Total landscape, theme parks, public space.* Aldershot, UK: Ashgate.

Molotch, H. (2002). Place in product. *International Journal of Urban and Regional Research, 26*(4), 665–688.

Morris, M.S. (1996). "Tha'lt be like a blush-rose when tha' grows up, my little lass": English cultural and gendered identity in *The Secret Garden. Environment and Planning D: Society and Space, 14*(1), 59–78.

Nakamura, L. (2008). *Digitizing race: Visual cultures of the Internet.* Minneapolis, MN: University of Minnesota Press.

Nelson, L., & Nelson, P.B. (2011). The global rural: Gentrification and linked migration in the rural USA. *Progress in Human Geography, 35*(4), 441–459.

Neumayer, C., & Raffl, C. (2008). Facebook for global protest: The potential and limits of social software for grassroots activism. *5th Prato Community Informatics & Development Informatics Conference 2008: ICTs for Social Inclusion: What is the Reality?*, 27 October-30 October 2008, Monash Centre, Prato, Italy.

Neuwirth, R. (2005). *Shadow cities: A billion squatters, a new urban world.* New York, NY: Routledge.

Nissenbaum, H. (2009). *Privacy in context: Technology, policy, and the integrity of social life.* Stanford, CA: Stanford University Press.

Novak, M. (1991). Liquid architectures in cyberspace. In M. Benedikt (Ed.), *Cyberspace* (pp. 225–254). Cambridge, MA: MIT Press.

Nunziato, D.C. (2005). The death of the public forum in cyberspace. *Berkeley Technology Law Journal, 20,* 1115–1171.

Olmsted, F.L. (1997 [1875]). 'Park' from the American encyclopedia. In C. Beveridge and C. Hoffman (Eds.), *The Papers of Frederick Law Olmsted* (pp. 308–329). Baltimore, MD: Johns Hopkins University Press.

Open data: A new goldmine. (2013). *The Economist.* May 18. Retrieved from: http://www.economist.com/news/business/21578084–making-official-data-public-could- spur-lots-innovation-new-goldmine.

Orbesen, J. (2013). When capitalism consumed the Internet. *Salon.* April 14. Retrieved from: http://www.salon.com/2013/04/14/when_did_the_internet_become_a_for_profit_venture _partner/.

Osenga, K. (2013). The Internet is not a super highway: Using metaphors to communicate information and communications policy. *Journal of Information Policy, 3,* 30–54.

Ostrom, E. (1990). *Governing the commons: The evolution of institutions for collective action.* Cambridge, MA: Cambridge University Press.

Ostrom, E., Burger, J., Field, C.B., Norgaard, R.B., & Policansky, D. (1999). Revisiting the commons: Local lessons, global challenges. *Science, 284*(5412), 278–282.

Pachucki, M.A., & Breiger, R.L. (2010). Cultural holes: Beyond relationality in social networks and culture. *Annual Review of Sociology, 36,* 205–224.

Papacharissi, Z. (2002). The virtual sphere: The Internet as a public sphere. *New Media Society, 4*(1), 9–27.

Papacharissi, Z. (2009). The virtual geographies of social networks: A comparative analysis of Facebook, LinkedIn and ASmallWorld. *New Media & Society, 11*(1–2), 199–220.

Pariser, E. (2011). *The filter bubble: What the Internet is hiding from you.* New York, NY: The Penguin Press.

Park, H.W., Barnett, G.A., & Chung, C.J. (2011). Structural changes in the 2003–2009 global hyperlink network. *Global networks, 11*(4), 522–542.

Pfeifle, M. (2009, July 6). A Nobel Peace Prize for Twitter? The free social-messaging utility uniquely documented and personalized the story of hope, heroism, and horror in Iran. *Christian Science Monitor.* July 6. Retrieved from: http://www.csmonitor.com/Commentary/Opinion/2009/0706/p09s02–coop.html.

Pine, J.B. (1921). *The story of Gramercy Park, 1831–1921.* New York, NY: Gramercy Park Association.

Poster, M. (1996). Cyberdemocracy: Internet and the public sphere. In: D. Porter (Ed.), *Internet culture* (pp. 201–218). New York, NY: Routledge.

Poster, M. (2006). *Information please: Culture and politics in the age of digital machines.* Durham, NC: Duke University Press.

Pow, C. (2007). Securing the 'civilized' enclaves: Gated communities and the moral geographies of exclusion in (post-)socialist Shanghai. *Urban Studies, 44*(8), 1539–1558.

Qualman, E. (2012). *Socialnomics: How social media transforms the way we live and do business.* Hoboken, NJ: John Wiley & Sons.

Raffel, S. (2013). *The method of metaphor.* Bristol, UK: Intellect Books.

Rangaswamy, N., & Cutrell, E. (2012). Re-sourceful networks: Notes from a mobile social networking platform in India. *Pacific Affairs, 85*(3), 587–606.

Rangaswamy, N., & Toyama, K. (2006). 'Global events local impacts': India's rural emerging markets. *Ethnographic Praxis in Industry Conference Proceedings, 1,* 198–213.

Rapoza, J. (2009). Social skills. *eWeek, 26*(10), 14–20.

Raymond, E. (1999). The cathedral and the bazaar. *Knowledge, Technology & Policy, 12*(3), 23–49.

Reichl, A.J. (1999). *Reconstructing Times Square: Politics and culture in urban development*. Lawrence, KS: University Press of Kansas.

Rheingold, H. (2000). *The virtual community: Homesteading on the electronic frontier*. Cambridge, MA: MIT Press.

Richardson, A.E., Montello, D.R., & Hegarty, M. (1999). Spatial knowledge acquisition from maps and from navigation in real and virtual environments. *Memory & Cognition, 27*(4), 741–750.

Roberts, J. (1998). English gardens in India. *Garden History, 26*(2), 115–135.

Roberts, J.M. (2001). Spatial governance and working class public spheres: The case of a chartist demonstration at Hyde Park. *Journal of Historical Sociology, 14*(3), 308–336.

Roberts, K. (2006). *Leisure in contemporary society*. Wallingford, UK: CABI.

Robinson, J. (2002). Global and world cities: A view from off the map. *International journal of urban and regional research, 26*(3), 531–554.

Rojek, C. (2009). *The labour of leisure: The culture of free time*. Thousand Oaks, CA: SAGE Publications.

Rosen, D., Barnett, G.A., & Kim, J.H. (2011). Social networks and online environments: When science and practice co-evolve. *Social Network Analysis and Mining, 1*(1), 27–42.

Rosen, G., & Razin, E. (2009). The rise of gated communities in Israel: Reflections on changing urban governance in a neo liberal era. *Urban Studies, 46*(8), 1702–1732.

Rosenzweig, R. (1979). Middle-class parks and working-class play: The struggle over recreational space in Worcester, Massachusetts, 1870–1910. *Radical History Review, 21*, 31–46.

Rosenzweig, R. (1985). *Eight hours for what we will: Workers and leisure in an industrial city, 1870–1920*. Cambridge, MA: Cambridge University Press.

Rosenzweig, R., & Blackmar, E. (1992). *The park and the people: A history of Central Park*. Ithaca, NY: Cornell University Press.

Rosol, M. (2010). Public participation in post-Fordist urban green space governance: The case of community gardens in Berlin. *International Journal of Urban and Regional Research, 34*(3), 548–563.

Roy, C. (2007). When wisdom speaks sparks fly: Raging grannies perform humor as protest. *Women's Studies Quarterly, 35*(3–4), 150–164.

Ruberg, B. (2006). Big reality: A chat with 'big game' designer Frank Lantz. *Gamasutra*. August 10. Retrieved from: http://www.gamasutra.com/view/feature/1847/big_reality_a_chat_with_big_.php.

Ruiz, N. (2008). Five social networking sites of the wealthy. *Forbes*. February 2. Retrieved from: http://www.forbes.com/2008/05/02/social-networks-vip-tech-personal- cx_nr_0502style.html.

Said, E. (1978). *Orientalism*. New York, NY: Pantheon.

Sardar, Z., & Ravetz, J.R. (1996). *Cyberfutures*. New York, NY: NYU Press.

Sassen, S. (2001). *The global city: New York, London, Tokyo*. Princeton, NJ: Princeton University Press.

Sassen, S. (2002a). Locating cities on global circuits. *Environment and Urbanization, 14*(1), 13–30.

Sassen, S. (2002b). Towards a sociology of information technology. *Current Sociology, 50*(3), 365–388.

Sassen, S. (2006a). Public interventions: The shifting meaning of the urban condition. *Open, 11*, 18–27.

Sassen, S. (2006b). *Territory, authority, rights: From medieval to global assemblages*. Princeton, NJ: Princeton University Press.

Saudis share almost 'everything' online. (2013). *Arab News*. June 3. Retrieved from: http://www.arabnews.com/news/453858.

Sawhney, H. (1992). The public telephone network: Stages in infrastructure development. *Telecommunications Policy, 16*(7), 538–552.

Sawhney, H. (2007). Strategies for increasing the conceptual yield of new technologies research. *Communication Monographs, 74*(3), 395–401.

Saxenian, A. (2006). *The new Argonauts: Regional advantage in a global economy*. Cambridge, MA: Harvard University Press.

Schmelzkopf, K. (2002). Incommensurability, land use, and the right to space: Community gardens in New York City. *Urban Geography, 23*(4), 323–343.

Scholz, T. (Ed.) (2012). *Digital labor: The Internet as playground and factory*. New York, NY: Routledge.

Scott, J. (1997). *Corporate business and capitalist classes*. New York: Oxford University Press.

Shi, M. (1998). From imperial gardens to public parks: The transformation of urban space in early twentieth-century Beijing. *Modern China, 24*, 219–54.

Shirky, C. (2008). *Here Comes Everybody. The Power of Organizing Without Organizations*. New York, NY: The Penguin Press.

Shirky, C. (2011). The political power of social media. Technology, the public sphere, and political change. *Foreign Affairs*. January/February. Retrieved from: http://www.foreignaffairs.com/articles/67038/clay-shirky/the-political-power-of-social-media.

Silva, A.D.S., & Hjorth, L. (2009). Playful urban spaces: A historical approach to mobile games. *Simulation & Gaming, 40*(5), 602–625.

Skeels, M.M., & Grudin, J. (2009). When social networks cross boundaries: A case study of workplace use of Facebook and LinkedIn. In: *Proceedings of the ACM 2009 international conference on Supporting group work* (pp. 95–104). New York, NY: ACM.

Smith, M. (2011). Contrasting Tea Party and Occupy Wall Street Twitter networks. *Connected Action*. November 16. Retrieved from: http://www.connectedaction.net/2011/11/16/contrasting-teaparty-and-occupywallstreet-twitter-networks/.

Snir, R., & Harpaz, I. (2002). Work–leisure relations: Leisure-orientation and the meaning of work. *Journal of Leisure Research, 34*(2), 178–203.

Soja, E.W. (2000). *Postmetropolis: Critical studies of cities and regions*. Oxford: Blackwell.

Solove, D.J. (2008). *Understanding privacy*. Cambridge, MA: Harvard University Press.

Sorkin, M. (1992). *Variations on a theme park: The new American city and the end of public space*. New York, NY: Farrar, Straus and Giroux.

Stalder, F. (2010). Digital commons. In: K. Hart, J-L. Laville, and A.D. Cattani (Eds.), *The human economy. A citizen's guide* (pp. 313–324). Cambridge, MA: Polity Press.

Stefik, M. (1996). *Internet dreams: Archetypes, myths, and metaphors*. Cambridge, MA: MIT Press.

Stutzman, F. (2006). An evaluation of identity-sharing behavior in social network communities. *Journal of the International Digital Media and Arts Association, 3*(1), 10- 18.

Sunstein, C. (2001). *Republic.com*. Princeton, NJ: Princeton University Press.

Takahashi, D. (2012). Social gaming pioneer Commagere pushes product placement as a moneymaker. *VentureBeat*. September 7. Retrieved from: http://venturebeat.com/2012/09/07/social-gaming-pioneer-blake-commageres-mediaspike-pushes-product-placement-as-a-moneymaker-for-social-games/#Pz4s1Hu1FEmspbKq.99.

Talbot, D. (2013). Facebook and Google create walled gardens for web newcomers overseas. *MIT Technology Review.* March 21. Retrieved from: http://www.technologyreview.com/news/512316/facebook-and-google-create-walled- gardens-for-web-newcomers-overseas/.

Tapscott, D. (2009). *Grown up digital: How the net generation is changing your world.* Columbus, OH: McGraw-Hill.

Tapscott, D., & Williams, A.D. (2006). *Wikinomics: How mass collaboration changes everything.* New York, NY: Penguin.

Taylor, A. (1995). 'Commons-stealers,' 'land-grabbers' and 'jerry-builders': Space, popular radicalism and the politics of public access in London, 1848–1880. *International Review of Social History, 40*(3), 383–407.

Taylor, P.J. (2004). *World city network: A global urban analysis.* London, UK: Routledge.

Taylor, T.L. (2003). Multiple pleasures women and online gaming. *Convergence: The International Journal of Research into New Media Technologies, 9*(1), 21–46.

Terranova, T. (2000). Free labor: Producing culture for the digital economy. *Social Text, 18*(2), 33–58.

Thompson, F. (1908, September). Amusing the Million. *Everybody's Magazine, 19,* 378–387.

Tonnies, F. (2002 [1887]). *Community and society.* Mineola, NY: Dover Publication.

Tranos, E., & Gillespie, A. (2011). The urban geography of Internet backbone networks in Europe: Roles and relations. *Journal of Urban Technology, 18*(1), 35–50.

Tseng, Y.F. (2011). Shanghai rush: Skilled migrants in a fantasy city. *Journal of Ethnic and Migration Studies, 37*(5), 765–784.

Turkle, S. (1994). Constructions and reconstructions of self in virtual reality: Playing in the MUDS. *Mind, Culture, and Activity, 1*(3), 158–167.

Turkle, S. (2012). *Alone together: Why we expect more from technology and less from each other.* New York, NY: Basic Books.

Turow, J. (2005). Audience construction and culture production: Marketing surveillance in the digital age. *Annals of the American Academy of Political and Social Science, 597*(1), 103–121.

Urry, J. (1995). *Consuming places.* London, UK: Routledge.

Vaidyanathan, G. (2008). Technology parks in a developing country: The case of India. *Journal of Technology Transfer, 33,* 285–299.

Van Den Hoven, J., & Weckert, J. (Eds.) (2008). *Information technology and moral philosophy.* Cambridge, MA: Cambridge University Press.

van Dijck, J., & Nieborg, D. (2009). Wikinomics and its discontents: A critical analysis of Web 2.0 business manifestos. *New Media & Society, 11*(5), 855–874.

Veblen, T. (1899). *The theory of the leisure class: An economic study in the evolution of institutions.* London, UK: Macmillan & Company.

Wajcman, J. (2008). Life in the fast lane? Towards a sociology of technology and time. *British Journal of Sociology, 59*(1), 59–77.

Wallis, C. (2011). New media practices in China: Youth patterns, processes, and politics. *International Journal of Communication, 5,* 406–436.

Warlaumont, H.G. (2010). Social networks and globalization: Facebook, YouTube and the impact of online communities on France's protectionist policies. *French Politics, 8*(2), 204–214.

Watkins, S.C. (2009). *The young and the digital: What the migration to social network sites, games, and anytime, anywhere media means for our future.* Boston, MA: Beacon Press.

Watson, A. (2012). The world according to iTunes: mapping urban networks of music production. *Global Networks, 12*(4), 446–466.

Weiss, J. (2006). *International handbook of virtual learning environments (vol. 2)*. New York, NY: Springer.

Westin, A.F. (2003). Social and political dimensions of privacy. *Journal of Social Issues, 59*(2), 431–453.

Williams, B. (2006). The paradox of parks. *Identities, 1*(1), 139–171.

Wilson, C., & Dunn, A. (2011). Digital media in the Egyptian revolution: Descriptive analysis from the Tahrir data sets. *International Journal of Communication, 5*, 1248–1272.

Wilson, M.I. (2001). Location location location: The geography of the dot com problem. *Environment and Planning B, 28*, 59–71.

Woody, T. (1957). Leisure in the light of history. *Annals of the American Academy of Political and Social Science, 313*(1), 4–10.

Woudstra, J., & Fieldhouse, K. (Eds.) (2000). *The regeneration of public parks*. London, UK: Spon Press.

Yang, G. (2003). The Internet and the rise of a transnational Chinese cultural sphere. *Media, Culture & Society, 25*(4), 469–490.

Yang, G. (2009). *The power of the internet in China: Citizen activism online*. New York, NY: Columbia University Press.

Yen, C.A. (2002). Western frontier or feudal society? Metaphors and perceptions of cyberspace. *Berkeley Technology Law Journal, 17*, 1207–1263.

Young, S. (2005). Morphings and ur-forms: From flâneur to driveur. *Scandinavian Journal of Media Arts Culture, 3*(3), 1–9.

Zhang, H., & Sonobe, T. (2011). Development of science and technology parks in China, 1988–2008. *Economics: The Open-Access, Open-Assessment E-Journal, 5*, 1–25.

Zhong, Y. (2012). The Chinese Internet: A separate closed monopoly board. *Journal of International Communication, 18*(1), 19–31.

Zook, M.A. (2003). Underground globalization: Mapping the space of flows of the Internet adult industry. *Environment and Planning A, 35*(7), 1261–1286.

Zoonen, L., van. (2005). *Entertaining the citizen: When politics and popular culture converge*. Lanham, MD: Rowman and Littlefield.

Zuckerman, E. (2008). The cute cat theory of digital activism. Paper presented at the E-Tech Conference, San Diego, CA, March. Retrieved from: http://en.oreilly.com/et2008/public/schedule/detail/1597.

Zukin, S. (1991). *Landscapes of power: From Detroit to Disney World*. Berkeley, CA: University of California Press.

Index

For Product Safety Concerns and Information please contact our EU
representative GPSR@taylorandfrancis.com
Taylor & Francis Verlag GmbH, Kaufingerstraße 24, 80331 München, Germany